ROPE SPLICING

ROPE SPLICING

BY
PERCY W. BLANDFORD

GLASGOW
BROWN, SON & FERGUSON, LTD., Publishers
4-10 Darnley Street

Copyright in all countries subject to the Berne Convention
All rights reserved

First Edition	–	1950
Second Edition	–	1969
Third Edition	–	1976
Fourth Edition	–	1990
Reprinted	–	1999

ISBN 0 85174 573 3

© 1999 Brown, Son & Ferguson, Ltd., Glasgow G41 2SD
Printed and Made in Great Britain

PREFACE TO THE THIRD EDITION

AS with the three earlier editions, this book is an attempt to gather together all kinds of fibre rope splices into one volume. While there are many excellent books on knotting and general ropework, which include splicing in their contents, this book is aimed at giving a more comprehensive coverage to splicing in particular.

In the years since the end of World War II, when the first edition was written, there has been a revolution in rope materials and construction due to the replacement of natural fibres by synthetic fibres. These have not responded to some traditional ways of splicing and a new range of splices has evolved to suit them. Splices for all kinds of rope are included in this book and it is believed that the reader will find more splices described than in any other single volume.

Only sufficient knotting and other supplementary information is included to make the descriptions self-contained. For information on knotting the reader is recommended to *Knots and Splices* by Capt. Jutsum, for fancy work to *Knots, Splices and Fancy Work* by C. L. Spencer or *The Harrison Book of Knots*, and for wire splicing to *Marline-Spike Seamanship* by Leonard Popple.

Some experienced readers may not agree with some of the names given to splices. Many splices have regional names, some have more than one name, and sometimes the same name has been given to more than one splice. No doubt, the originators were more concerned with uses than names. In this book those names which the author has found most commonly applied are used.

It is impossible to acknowledge all the help given in supplying information for the compilation of this book, but the Samson Cordage Company, of Boston, Massachusetts, were particularly helpful with information on the splicing of modern braided cordage, while the author is particularly appreciative of the information provided by the British firm of Marlow Ropes Ltd.

Percy W. Blandford.

CONTENTS

CHAPTER 1
Splicing Preliminaries—Natural ropes—Synthetic ropes—Tools ... 9

CHAPTER 2
Basic Eye Splice 16

CHAPTER 3
Basic Joining and End Splices—Back Splice—Short or butt splice—Long Splice 24

CHAPTER 4
Other Eye Splices—Branch splice—Cut and log line splices—Eye in middle of rope—Sailmaker's splice—Eye Splice, wormed and collared—Eye Splice with collar—Eye splice, grafted—German eye splice—Chain splice—Grommet splice—Flemish eye 31

CHAPTER 5
End-to End Splices—Shroud knots—Short splices (three-strand into four strand)—Long splice (three-strand into four-strand)—Short splice, knotted—Short splice (rope to wire)—Long splice (rope to wire)—Grecian splice 47

CHAPTER 6
Splicing Cables—Ropemaker's eye—Admiral Elliott's eye—Common eye splice (cable)—Short splice (cable)—Long splice (cable)—Staggered short splice (cable) 61

CHAPTER 7
Braided Rope Eye Splices—Tools—Flemish eyes—Lock tuck splice—Knights eye splice—Samson eye splice 66

CHAPTER 8
Other Braided Rope Splices—Back splice—Single braid end-to-end splice—Double braid end-to-end splice—Joining thin and thick ropes—Wire to fibre rope splice 82

CHAPTER 9
Miscellaneous Splices—Eye splice (multi-plait rope)—Two ended eye splice—Cringle to sail—Cringle to rope—Grommet—Selvagee strop—Flag rope—Pudding splice—Cockscombing 93

Glossary and Index 105

ROPE SPLICING

CHAPTER 1

SPLICING PRELIMINARIES

A SPLICE is a more permanent piece of ropework than a knot or bend. It is made by using the strands or yarns of the rope, which are interwoven. Where nearly all knots are made with the whole rope, a splice uses parts of the rope. A splice is normally stronger than knots which might be used for the same purpose. It should be safer as there is less risk of it coming apart. There are a few ornamental knots that come between splices and practical knots, as they use strands of the rope. Where they have uses comparable to splices as well as being decorative, they are treated in this book as splices.

Natural ropes.

For most of the thousands of years that man has made rope, he has used natural materials, but in the years following World War II synthetic materials have taken over many of the uses of natural materials, so the majority of ropes used at sea and on shore are now formed of man-made fibres. These have many advantages and few disadvantages, but because of their nature they require special splicing techniques.

Rope has been made of just about every fibrous material, as well as hair. In recent times the common natural materials used for rope have been: hemp, manila, flax, sisal, coir and cotton. Hemp and manila will hold their shape when strands are opened for splicing. For yacht ropes and other purposes where strength is required, these are the natural fibre ropes to choose. Flax and sisal are rougher, but should be easy to splice. Coir has little strength, but it floats, hence its use for towing and mooring. New cotton is white, but soon gets dirty. Neat splices in it are difficult as it does not always keep its shape when opened.

Synthetic ropes.

The fibres used in natural ropes are all comparatively short. When made up into rope the ends cause the roughness or hairiness of the feel when the rope is handled. This had advantages in providing a grip and it helps the tucks of a splice to hold. Instead of fibres a synthetic rope is made up of continuous filaments. These may be extremely fine, with a very large number made up into each yarn or strand. There may be occasional joins, but in general each filament can be assumed to go the full length of the rope. This means that a synthetic rope may be very smooth, which makes for easy running, but the lack of hairiness reduces the grip, both for handling and splicing. Because of this some synthetic ropes are made with ends projecting to give hairiness (spun or stapled), so it is unwise to assume that every hairy rope is natural fibre, although every smooth rope should be synthetic.

A great many synthetic materials are used for making ropes. Others may be introduced, but splicing methods are likely to be equally suitable. The general-purpose rope of good strength, used for halliards and any situation where stress has to be taken without stretch, is polyester (Terylene, Dacron). Other non-stretch synthetics may be substituted for it, mainly for economy. Nylon is slightly stronger than polyester, but it differs from most other synthetics in being elastic and capable of absorbing water. Its elasticity makes it unsuitable for situations where an unyielding pull is needed, as in halliards and most sheets on a boat, or in a crane hoist. It is used afloat for towing, mooring and anchoring. Its water absorption is slight and of little consequence.

Some other synthetics are derived from propylene, a gaseous hydrocarbon. Polypropylene is one of these. It looks like nylon when made into fine filaments. It is only about half the strength of nylon, but it will float. Polyethylene is another of this group, which is made into an economical rope, which is usually coarse and stiff. However, it is difficult to identify the different synthetic ropes as they can be made tight or loose, coarse or fine. Some makers include a coloured thread for identification.

Synthetic fibre ropes are all stronger than their natural fibre counterparts. They keep their strength better, as they are immune to the effects of damp. Many are better able to resist abrasion. Damp is the enemy of natural fibre ropes as it gets trapped inside and causes rot. If natural fibre rope is opened and seen to be black at the centre, rot has taken a hold and weakened the rope.

Because synthetic ropes are stronger than natural fibre ropes, a thinner one may be used, but there are the practical considerations of handling. Anything less than 9mm diameter is difficult to grip. Traditionally, rope sizes have been given as their circumference in inches. Metric sizes are given as millimetres diameter. There is a convenient way of converting: take the circumference in eighth-inches and call it millimetres diameter. For example: 1⅛in circumference is 9 eighths inches, so the diameter is 9mm.

Heat will weaken ropes, so chafe has to be controlled, but excessive heat will actually melt synthetic ropes. This provides a means of identification. If a flame is put to the end of a natural fibre rope it will char and burn. Synthetic ropes will melt first and may only burn after melting. Melting the ends of the synthetic filaments together is a convenient way of sealing the end of the rope temporarily. There are cutting devices that use a heated knife to both cut and seal the severed ends.

There are now many ways of making up rope, but the traditional three-strand rope is still most popular, in synthetic as well as natural fibres. Most rope has the three strands laid up right-handed. If you look along the rope, the strands spiral away from you in a clockwise direction. There are left-handed ropes, laid up the other way. Using right-handed and left-handed ropes for slings reduces the risk of twisting. The term 'hawser-laid' may be met as a name for three-strand rope.

There are four-strand ropes (sometimes called 'shroud-laid'). In cross-section four strands do not touch each other the same as three strands will, so in larger ropes there may be a smaller central straight strand, put there to keep the rope in shape.

In a rope the fibres are twisted together one way to make yarns, which are twisted the opposite way to make strands, which are then put together

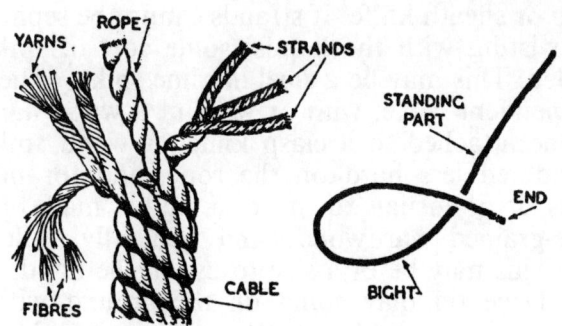

Fig. 1—The parts of a rope

the other way to make the three-strand rope. For a very large cable, three ropes may be 'cable-laid' together. Traditional terms applicable to ropework are shown in Fig. 1.

Besides strand ropes there are plaited or braided ropes. The outside of the rope is made up of a sort of round plait with fibres woven into each other. There may be a second braided sheath inside the first. The centre may be straight strands to make up the size or the strands may be laid up like a three-strand rope. Braided rope is easy to handle and is not so liable to kink as strand rope. It is being used increasingly on yachts and elsewhere. Normal splices cannot be used and some special ones have been devised.(*see* Chapter 7)

Tools.

For many splices the tools needed are few. Some work can be done with the hands only. A sharp knife is needed for trimming and one with replaceable blades may now take the place of the traditional clasp or sheath knife. If strands cannot be separated by twisting with the hands, some sort of spike is needed. This may be a steel marline spike, either an independent spike, with or without a wood handle, or one attached to a clasp knife. A wood spike is considered less hard on the rope, and for bigger ropes it is usual to have a spike made from close-grained hardwood, and generally called a 'fid'. This may be of a size to use in the hand, or a very large fid may stand on the ground with its point upwards, so the rope has to be pressed on it.

A traditional spike or fid is used to force strands apart, then it is withdrawn and the tucking strand entered before the gap closes. Many synthetic ropes will close so quickly that this cannot be done. One way to hold the gap open is to use a broad screwdriver, with any sharp edges removed, instead of a spike. This can be pushed in, then turned on edge so the tucking end strand can be entered beside it. There are spikes with broad flat ends to use in a similar way, but a better tool is usually described as a 'Swedish fid'. This has a tapered hollowed sheet metal end on a wood handle. When pushed between the main rope strands and turned on edge there is a more spacious hollow to admit the tucking end, before the fid is withdrawn.

Modern types of braided rope need new splices and these require some special tools in addition to the traditional one (*see* page 67).

Grit is an enemy of rope, so keep rope being worked on off the floor. The practice of rolling a splice underfoot to press it into a neat shape may not have been harmful on a ship in the middle of an ocean away from shore dirt, but ashore it could mean sharp grit being forced into the rope to cut and weaken it.

CHAPTER 2

BASIC EYE SPLICE

SPLICING strand rope involves tucking end strands over and under their parts in the body of the rope. Differences come in the way the first rounds of tucks are made, but after that most splices are made by tucking alternately over and under main strands. The most used splice is the eye splice in the end of a rope. This may be a simple loop or one made tightly around a thimble, which protects it from chafe when attached to a shackle or other metal fitting. It could also be made tightly in the same way around a grooved piece of wood forming a toggle on a flag.

For an eye of no rigidly-fixed size, first unlay the strands in the end of the rope for a sufficient distance. This is usually a length greater than will be used up in the tucks, to give some spare for convenience in handling. For rope 12mm ($\frac{1}{2}$in) diameter or less, about 20cm (8in) should do. Some rope will hold together without special treatment, but most synthetics should have the end of each strand sealed. Heat with a cigarette lighter or match until the end softens. Moisten your finger and thumb and roll the end to a rounded point. It will set almost immediately. Some ropes tend to unlay themselves once they are opened and are difficult to lay up again neatly.

ROPE SPLICING

To prevent this, put on a few turns of whipping line temporarily at the selected distance from the end before separating the strands. A West

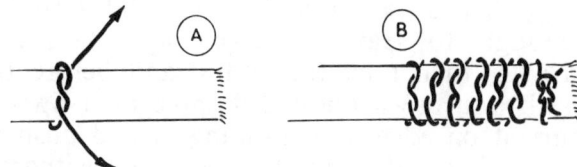

FIG. 2—A West Country whipping

Country whipping is convenient for this and other whippings associated with splices. Overhand knots are tied alternately back and front (Fig. 2A). After covering a length no more than equal to the thickness of the rope, the last knot is made into a reef knot (Fig. 2B).

Bend the rope into the size loop required and arrange the ends so two are across the lay and one is out of the way behind (Fig. 3A). Lift one

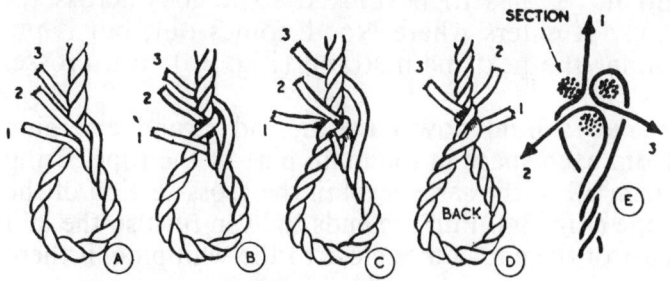

FIG. 3—An eye splice in three-stranded rope

main strand. Sometimes the rope can be twisted open enough in the hands, otherwise open the space with a spike. Be careful that the spike only goes between the strands and does not push into a strand and break the fibres or filaments. Push No. 2 end strand under this lifted main strand (Fig. 3B). If you use a round spike or fid, push it in far enough to leave a large gap when it is withdrawn, then tuck the end strand before that closes. If you use a flat-ended spike or a Swedish fid, turn it on edge after pushing in and push the tucking end strand along its side before withdrawing it.

Keep No. 1 strand on the loop side of No. 2. Note where No. 2 strand comes out—No. 1 strand has to go in there. Lift the next main strand and tuck No. 1 strand under it (Fig. 3C). Check that the two end strands emerge from adjoining spaces.

Turn the splice over and find the main strand that does not have an end strand under it. No. 3 strand has to go under it, but be careful how you do it. It has to be tucked so it goes across the lay. It enters where No. 1 comes out, but is put under the next main strand (Fig. 3D) so it crosses its lay.

You should now have an end strand emerging from each space in the main part of the rope. Bring them all to the same level in the cross-section of the rope (Fig. 3E). Pull the ends tight in turn so the laid part of the end comes close to the whipping, if there is one.

Take any end strand and tuck it over the adjoining main strand and under the next, working across the lay—the main strands are clockwise in the right-handed rope shown, so you tuck anticlockwise. Having got that end strand tucked, move to the next one and do the same on the next main strand around the rope. Then do it with the third end and main strands. Each end strand will now have gone under and over twice and you will have an end emerging from each space again.

A common fault is to tuck too far along the rope. The tucked ends should go around the rope at about the same angle as the main strands, but in the opposite directions. Work the tucked strands in turn down towards the eye if your splice is becoming 'long-jawed'.

Tuck each strand in the same way once more. This is called 'three whole tucks' (Fig. 4). In natural fibre ropes this is enough to provide

Fig. 4—An eye splice with three whole tucks

strength. In synthetic ropes it is better to make five whole tucks, to provide adequate friction between the much smoother strands.

If a splice is left with just the whole tucks, it does not finish very neatly and there is a rather abrupt change of section. It is better to give the end of the splice a taper. With a knife scrape away about one-third of the fibres or filaments of each end strand and tuck once more. After all three ends have been tucked, scrape away half of the remainder and tuck again. Such a splice would be described as having 'three (or five) whole and two taper tucks'.

With natural fibre rope the ends of the strands can be cut off fairly close to the splice and left. With synthetic rope it is better to seal the ends. First cut the ends within about 5mm ($\frac{1}{4}$in) of the splice, then use a flame to melt the ends of the filaments together. Be careful not to let the heat reach and soften the splice of the rope.

A complete splice sometimes looks uneven. It is difficult to get exactly the same tension and angle on each tuck. Instead of risking grit entering if the splice is rolled under foot it is better to roll it between two boards or under a board on a bench top. It is also advisable to put some tension on the splice. This will bed down the tucks. If much unevenness has to be corrected, this is better done before cutting off the ends.

No splice can be as strong as the rope, but a well-made splice is stronger than the best knot for the same purpose. An eye splice, as just described is about 90% as strong as the rope, while a bowline knot is less than 80%.

Where the utmost strength is required and appearance is of secondary importance, there can be six whole tucks in natural fibre rope and eight whole tucks in synthetic rope. There are no taper tucks, but the ends are 'dog-knotted' together. Each end is halved. The halves are seized to their neighbours (Fig. 5). If there is a sufficient length left, it may be possible to pull fibres or filaments from the ends and twist them into whipping line to make the seizing, but it is more likely that separate whipping line, preferably of the same material as the rope, will be used. A

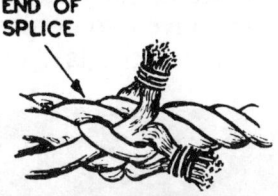

Fig. 5—Dog knots

West Country whipping makes a suitable seizing. Use the two ends of the line around the dogged ends and make tight overhand knots back and front alternately for about ten times, then make the last knot into a reef knot.

If the eye splice has to be made around a thimble, the method of tucking is the same, but the first tucks have to be located so as to finish tight enough to hold the thimble. Prepare the rope as described. Wrap around the thimble, with the whipping or the point where the end strands separate close to the main part of the rope at the point of the thimble. Tuck No. 2 strand under the nearest convenient main strand close to the thimble. Check that this will make a tight fit when pulled through, then slacken enough to get the thimble out. Tuck No. 1 and No. 3 strands as before.

Put the thimble back in the loop and draw the end strands tight (Fig. 6). There can be some adjustment of grip on the thimble by moving these tucks up or down the rope slightly. When a satisfactory fit has been achieved, leave the thimble in place and continue to make the other whole and taper tucks to complete the splice.

An eye splice is often left as described, but it can be made neater and given some protection if it is served over, at least for the last few tucks. It is

Fig. 6—An eye splice around a thimble, with all ends tucked once

advisable to tension the splice first. By putting it under load for a short time the tucked strands will settle down so the seizing will stay put better. The seizing should be whipping line or something slightly thicker, preferably of the same material as the spliced rope. Turns are put on as tightly as possible. They can be made in the same way as a common whipping, with the ends under the turns, or the West Country whipping can be used.

ROPE SPLICING

Cover the end tucks and work back towards the eye, but the first one or two tucks can be left exposed.

CHAPTER 3

BASIC JOINING AND END SPLICES

ALTHOUGH the eye splice is the one most frequently used, the back splice and short splice are often needed, while the long splice has its own advantages. These are the main four working splices in strand rope. Beside the practical advantages of learning to make them, they form the basis of many other splices.

Back splice.

For most purposes the end of a rope is whipped. If it is a synthetic rope it may also be sealed by heating and melting the ends of the filaments and strands together. An alternative is to use the strands of the rope itself to make a back splice. This has the advantage of not needing anything else, but it has the disadvantage of thickening the end of the rope, so it may be difficult to pass through an eye or a block. A back splice should be more secure than a whipping, so if the greater thickness does not matter, it may be preferable.

Start by opening the strands with enough to handle and allow for tucking. As with an eye splice, a minimum of about 20cm (8in) should do. If necessary, seal the ends of the strands, and put a few temporary turns of whipping line around the point where the strands separate.

ROPE SPLICING

Hold the rope in one hand with the strands evenly-spaced (Fig. 7A). For the usual right-handed rope, take each strand in turn anti-

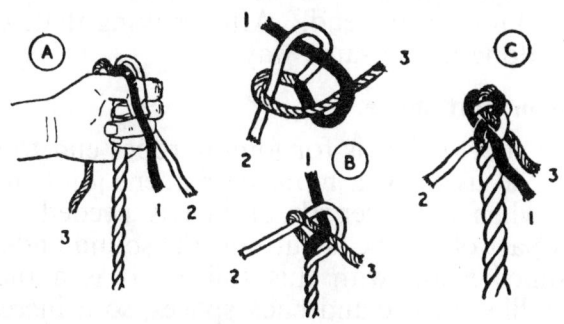

Fig. 7—A back splice

clockwise (when viewed from above) over its neighbour—1 over 2, 2 over 3 and 3 over 1 and down through the loop already there (Fig. 7B). This is a crown knot. Pull tight, so the pattern of the knot is even and flat across the end and each end points down and across the lay of the rope. If a temporary whipping has been used, remove it.

The ends will be seen to be pointing the right way for tucking over and under in the same way as in the completion of an eye splice. Lift a main strand, using a spike if necessary. Take an end strand over the main strand it adjoins and under the lifted one (Fig. 7C). Do this with the other ends. Repeat tucking each end over and under one main strand in turn. There could be more tucks, but as this is not a load-bearing splice, it will probably be sufficient for most purposes to now taper each strand and tuck it once or twice more.

Four-strand rope can be back-spliced in the same way. The only difference is in the four-part crown knot on the end. After making that, tucking is done in the same way.

Short or butt splice.

The short splice is for joining ropes end to end. It can be used for a more permanent joint than a knot when extra length of line is needed. If a worn part of a rope is cut out, the sound ends can be joined again with this splice. It is a tucked splice, like the eye and back splices, so it increases the thickness of the rope. It is economical in material, but if the joined rope has to pass through a block or around a wheel, the more wasteful long splice must be used.

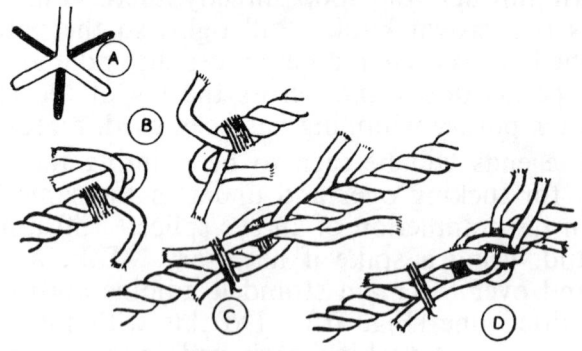

FIG. 8—A short splice

Unlay a working length of each rope, seal the ends if necessary and put whippings temporarily at the points where the strands are unlaid. Even with rope which does not have a tendency to unlay accidentally, whippings are advisable when learning this splice. Bring the ends together so each end comes opposite a space (Fig. 8A). Push the ropes tightly together (Fig. 8B) so the whippings are as close as possible.

Splicing is done one way at a time, so it may be helpful to temporarily seize down the strands in the direction not yet being used (Fig. 8C). With natural fibre rope the whipping in the direction being tucked may be removed. With synthetic fibre rope it is usually better to leave the whipping in place until the opposing ends are loosely tucked, then remove the whipping before tightening them. This prevents the lay of the rope opening and becoming uneven.

Tuck each free strand over and under one of the strands of the other rope (Fig. 8D). For natural fibre ropes make two whole tucks in this way. For synthetic fibre ropes make four whole tucks. Follow with taper tucks for neatness or dog knot the ends together if the maximum strength is required. Remove the temporary seizing around the other end strands and repeat the process the other way. See that the tucks near the centre of the splice are kept close together so the final effect is an appearance of continuous tucking.

Long splice.

The long splice joins ropes without any appreciable thickening of them, so it can follow the rope through blocks or around a windlass without causing any obstruction or jamming. When properly made, it is difficult to distinguish the splice from the rest of the rope. Against that is the fact that it uses up a considerable amount of rope. How much of each rope to be unlaid depends on the degree of strength required and the material from which the rope is made.

For natural fibre three-strand ropes the absolute minimum to be unlaid is 7 times the circumference (22 times the diameter) and 10 times the circumference (32 times the diameter) for four-strand natural fibre ropes. This is better increaesd to 12 or 14 times the circumference (38 or 45 times the diameter) for three-strand and half as much again for four-strand.

For synthetic fibre ropes, this should be increased by at least a further 50%. As an example, a 12mm ($\frac{1}{2}$in) diameter synthetic fibre rope will require both ends to be unlaid at least 72cm (29in). As both ends use up this amount, the length of rope consumed will be about 1·50m or 5ft.

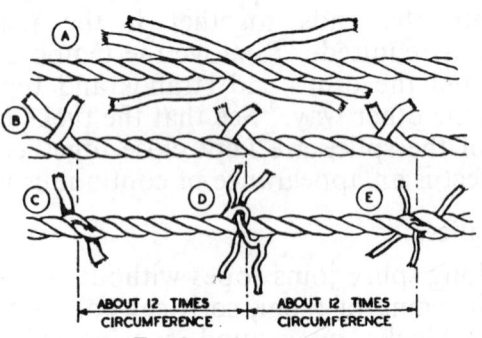

FIG. 9—A long splice

ROPE SPLICING

Unlay the ropes for the required distance and bring them together as if for a short splice (Fig. 9A). Without disturbing the other strands, start to unlay one strand further and as you do this lay in its adjoining end strand from the other rope. Do not twist the strand being laid, but fit it in the space exactly as the one that is being removed. Continue doing this until you are within a short distance of the end of the strand. The distance shown in the example applies to natural fibre three strand rope. Take another pair and work the same manner the opposite way. Let the remaining pair of strands stay put (Fig. 9B).

The ends of the strands now have to be locked in some way, and a variety of methods are possible, among them are the following:

(i) Tuck each end under its adjoining strand, then take it around and under the same strand three times, tapering well for the last two tucks.

(ii) Divide the yarns in each strand into two equal groups. Tuck each half-strand around a separate main strand: take one half under and around the adjoining strand and the other half over the adjoining strand and under and around the next one.

(iii) Half-knot the two ends together (Fig. 9c), then tuck as in a short splice, tapering well after the first tuck.

(iv) Divide the yarns in each strand into two equal groups. Half-knot two of the half-strands together (Fig. 9D). Tuck the other half-strands under the adjoining main strands (Fig. 9E) and continue, after tapering, tucking it around that strand. Taper the knotted half-strands, take them over the adjoining strands and under and around the next strands.

Splicing four-strand rope is similar, but the greater length is needed to allow even spacing of the joined strands. Bring the rope ends together and work with one pair of strands, unlaying one and putting the other in its place as far as possible in one direction, allowing a little for tucking. Do the same with another pair in the opposite direction. Measure the distance between these two extreme positions. Divide this by three and unlay and lay the other pairs of strands to these one-third distances. This gives the meeting positions an equal spacing in the length of the splice. Treat all of the meeting ends in one of the ways described for the three-strand splice.

Before finally cutting off and sealing ends, tension the splice, so the strands settle in their new positions and the joints become as compact as possible. Roll the joints if necessary. Trim the tucked ends close and the splice should become inconspicuous after the rope has had a little use.

CHAPTER 4

OTHER EYE SPLICES

WHILE the basic eye splice in three-strand rope serves many purposes, there are occasions when variations on it or other means of making an eye are needed.

A normal eye splice in four-strand rope is similar to one in three-strand rope, once the problem of tucking the extra strand has been dealt with. Start as for a three-strand splice, but have three strands in front pointing across the lay

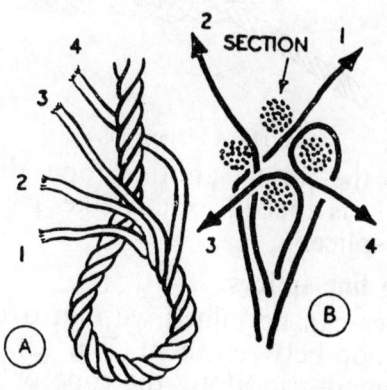

FIG. 10—An eye splice in four-stranded rope

and the fourth at the back (Fig. 10A). Tuck end 3 under the most convenient main strand. Tuck end 2 in where end 3 comes out and under the next main strand. Tuck end 1 in the same place as end 2, but go under two main strands. Turn the splice over and tuck end 4 under the remaining main strand, going in where end 1 comes out (Fig. 10B). It will be seen that ends 2, 3 and 4 are tucked in the same way as for a three-strand splice. If tucked correctly, there will be an end strand emerging from each space in the rope. From this point continue tucking in the same way as for a three-strand splice.

Branch splice.

If rope has to be spliced to give a Y formation, as in a divided tow rope, or one rope has to be

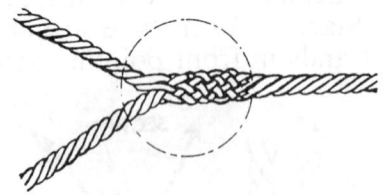

FIG. 11—A branch splice

spliced into the side of another rope, the splice to use (Fig. 11) is tucked exactly like a basic three-strand eye splice.

Cut and log line splices.

Two ropes can be joined with an overlap so as to leave a loop between them (Fig. 12). This may provide a hand-grip along the rope or it is useful to slip a line over the end of a tiller or for use in the shrouds of a dinghy's mast. The two sides of the loop need not be the same length—one long side may be useful for holding or for a continuous

ROPE SPLICING

rope slung over posts to form a rail. This version may be called a horseshoe splice.

To make a cut splice, unlay sufficient of each

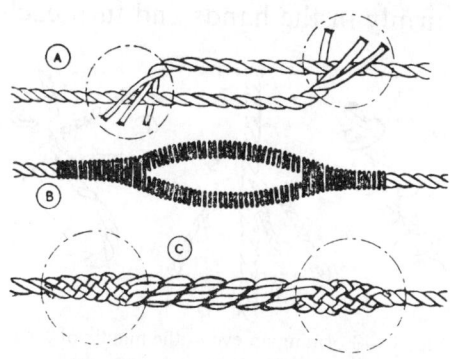

Fig. 12—Cut and log line splices

end, then overlap the ends for the amount needed for the loop. Lay the end strands over the main parts exactly as for making an eye splice (Fig. 12A). Tuck as for eye splices. If the loop has to fit over a solid object it can be served over to provide protection (Fig. 12B).

The log line splice is a variation (Fig. 12C), which can be used as an alternative to a short splice for joining ropes. It can also provide a thickening to serve as a grip or an indication of the amount of rope let out when the line slides through the hand. Start as for a cut splice. Make the tucks at one end, then twist the ropes together in the opposite way to their lay, and make the tucks at the other end.

Eye in middle of rope.

This is an ordinary eye splice, except that

instead of single strands, the tucks are made with twisted double strands.

To make the twisted strands, take a portion of the rope firmly in the hands and turn each hand in

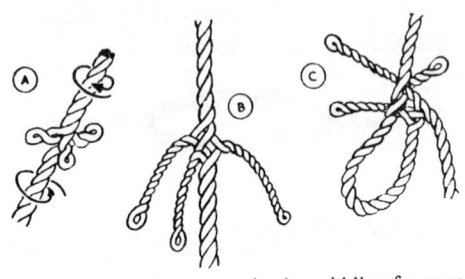

FIG. 13—Forming an eye in the middle of a rope

opposite directions so that there is a tendency for the natural twist of the strands to straighten out (Fig. 13A). As the turning is continued, the strands will separate and continue twisting back around themselves. Continue in this way until the three 'ends' are long enough for tucking (Fig. 13B). Form the rope into an eye, then commence tucking with the 'ends' exactly as for a common eye splice (Fig. 13C).

Sailmaker's splice.

When an eye has to be formed in the end of a bolt rope of a sail, it helps in sewing the sail to the rope to keep the form and lay of the rope right into the splice instead of the usual woven pattern of tucking. This type of splice is not as strong as a basic eye splice, and it would not do for a load-bearing eye in a free rope, but it has adequate

ROPE SPLICING

strength for its purpose. It should only be used for a rope that is sewn into a sail.

Prepare the rope ends for tucking, but put the eye together the opposite way to the start of a basic eye splice, with the front ends pointing in the direction of the lay, instead of across it (Fig. 14A), and the third end at the back. Tuck end 2 (Fig. 14B) under a convenient main strand. Tuck

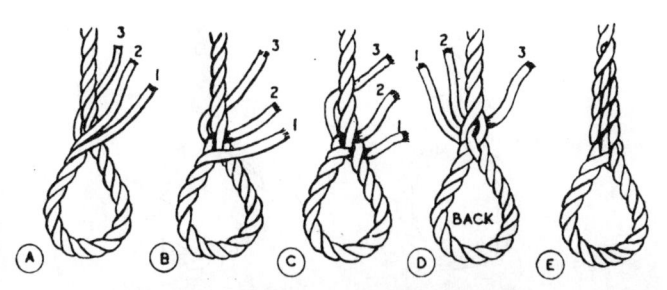

FIG. 14—A sailmaker's splice (first method)

end 1 where end 2 comes out and under the next main strand (Fig. 14C). Turn the splice over and locate the main strand with nothing under it. Tuck end 3 there (Fig. 14D). The sequence of tucks will be seen to be the same as for a basic eye splice, but they are in the opposite direction.

Check that an end strand emerges from each space. From this point the end does not go over and under, but is wrapped back around the strand it is already under. This is continued for a sufficient length (Fig. 14E). For neatness it is usual to start tapering after only one or two tucks. For natural fibre rope four tucks should be sufficient, but for synthetic fibre rope two more would be advisable. In this case, a tuck is the number of times the end goes around a main strand.

An alternative, but less satisfactory method, is started by first tucking as for a basic eye splice. After the three ends have been tucked (Fig. 15A), each strand is wrapped back around the main strand it is already under. This results in the

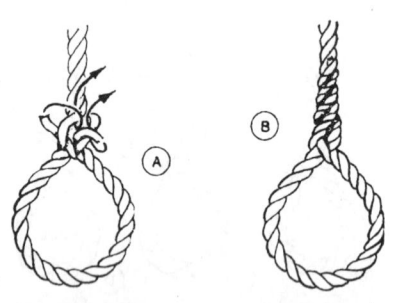

FIG. 15—A sailmaker's splice (second method)

turns around main strands being in the opposite direction to those of the first method (Fig. 15B). The result is not as neat and may not be as strong. To get the twist the other way after starting in this way, each end can be taken around the main strand at its other side instead of the one it is under, but this is not as neat and satisfactory as the first method.

Eye splice, wormed and collared.

This is a purely ornamental variation of the sailmaker's eye splice, which looks most effective in four-strand rope. It is longer and uses up more rope than the ordinary splice. Because of the

ROPE SPLICING

decorative work at the end of the splice it cannot be run close up to a block.

To make the splice, commence by making the first tuck as if starting a sailmaker's splice, but using considerably longer strands (Fig. 16A). From

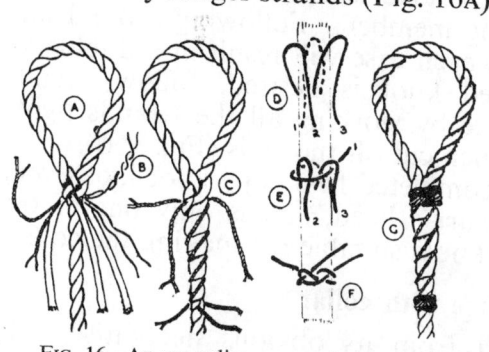

FIG. 16—An eye splice, wormed and collared

each strand separate four yarns and with the remaining yarns complete a sailmaker's splice, tapering it well.

Lay up the remaining yarns into two-yarn nettles, i.e. twist them up in pairs between finger and thumb left-handed (Fig. 16B), making two nettles for each strand.

Take one nettle from each strand and worm it around the rope (i.e. lay it in a space between two strands) down to the end of the splice, then tuck it under one strand of the rope. Arrange all the nettles so that they finish level (Fig. 16C).

The two groups of nettles are each formed into footrope knots, in the following way: Take each nettle and double it back along the rope, keeping the end to the right of its fixed part (Fig. 16D).

ROPE SPLICING

Working to the right, take one end around both parts of its neighbour and through the next bight (Fig. 16E). Do the same with the other ends. This completes a single footrope knot (Fig. 16F). To double it take each end around alongside its adjoining member. Following around under two parts in each case will result in a double knot. If a trebled knot is desired, follow round again. Complete by working all the strands tight with a spike, then cut off the ends (Fig. 16G).

The completed foot rope knot looks very much like a Turk's head, and is only one of the ornamental knots suitable for making the collars.

Eye splice with collar.

Apart from its obvious decorative value this splice is useful where the eye has to run up close to a block.

The splice is started as for a common eye splice, but each strand is only tucked once, then hove

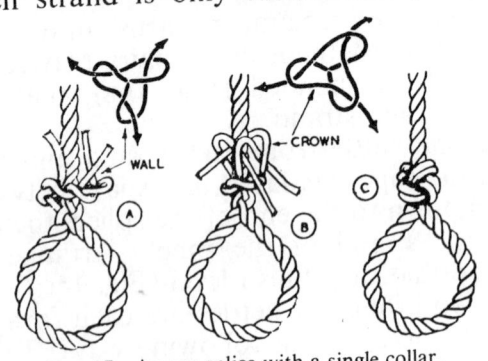

Fig. 17—An eye splice with a single collar

tight. The strands are then worked into a collar

which may be a footrope knot, as just described, or a manrope knot.

A manrope knot, which consists of a wall knot, topped by a crown knot and followed round once, is made as follows:

Take each strand to the right and pass it up under its neighbour, the last strand going up through the bight of the first (Fig. 17A). Continuing the same way round, take each strand down over its neighbour (Fig. 17B). This is shown in greater detail in Fig. 7. Now let each strand follow around again, keeping alongside the strand it is already adjoining. Work as tightly as possible and cut the strands off short (Fig. 17C).

Eye splice, grafted.

'Grafting' is the name given to the interweaving of the yarns of a rope so as to obtain an ornamental finish. The process is the same as in one method of 'pointing'. Splices may be completed by grafting, one example only being given here. A large number of patterns are possible, but the reader is referred to a book on fancy ropework for further information on this subject.

An eye splice may be grafted in two ways: the strands may be tucked once and the strands separated into yarns which are used for grafting (Fig. 18A), or the splice may be completed and separate strands seized on to make the grafting. Whichever method is used, work sufficient yarns around the rope as worming then arrange the remaining yarns into an even number of two-yarn nettles. Take alternate nettles along the rope

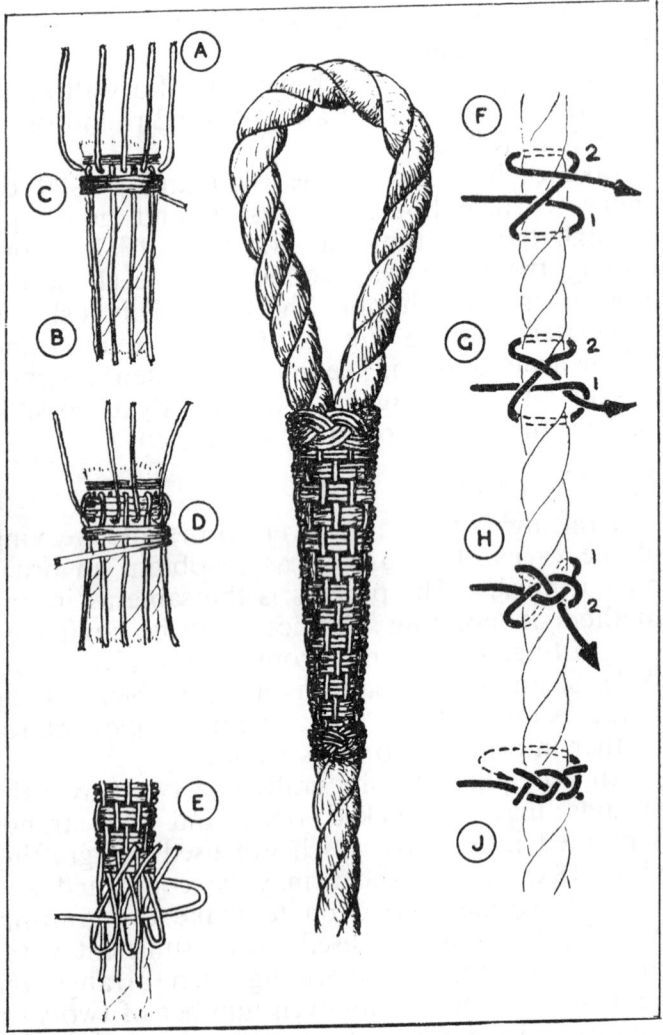

FIG. 18—An eye splice, grafted

(Fig. 18B). Half-hitch two or three turns of light line around the under nettles close up to their junction. This is called the warp (Fig. 18C). Lay the turned-back nettles down along the rope and turn back the others (Fig. 18D). Half-hitch on two or three more turns of warp. Continue this up and down weaving for a sufficient length. Finish by turning back the ends of one set of nettles and threading the warp through them (Fig. 18E). A neater effect is obtained if a few nettles are used to form footrope knots at the beginning and end of the grafting.

While footrope knots are best, an easier alternative is to add separate Turk's head knots, made in the following way:

Take the line around, almost the same way as in making a clove hitch, except that the end does not go under the last turn (Fig. 18F). Take this end under the first turn (Fig. 18G) and lift the second turn over the first (Fig. 18H). Pass the end down between the loops (Fig. 18J). This completes a single Turk's head. Double or treble by following around again.

German eye splice.

There are variations on the basic splice that make the first tucks in different ways. The important thing is for the ends to project one from each space ready for further over and under tucking. In some forms of wire splicing there is a locking tuck, where two ends go opposite ways around a main strand. It is unusual to find a locking tuck in fibre rope splicing, but it is used in the German eye splice, although it seems unlikely that it offers any advantages.

Ends are prepared and all three laid across the lay of the main strands (Fig. 19A). Tuck the strand furthest from the eye under a convenient main strand (Fig. 19B). Tuck end 2 below it and

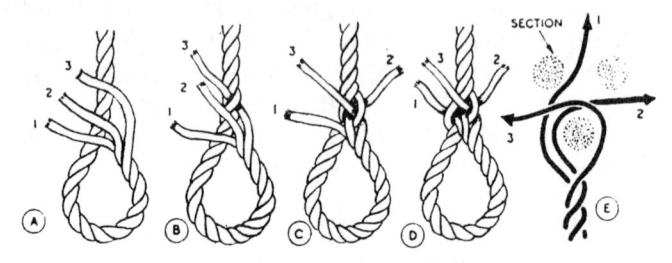

FIG. 19—A German eye splice

under the same strand the opposite way (Fig. 19C). Take end 1 in the same space, but under the next strand (Fig. 19D). This should leave one end emerging from each space (Fig. 19E). From this point tuck over and under one against the lay, as for a basic eye splice.

Chain splice.

A problem comes when a rope has to be spliced to the end of a chain in such a way that the rope will follow the chain through the narrow space of a fairlead or hawse pipe, where there may not be enough clearance to pass the double thickness of a normal eye splice. A chain splice uses only two of the three rope strands through the end of the chain and so reduces bulk. It also reduces strength, so this splice is only of use where the rope has to merely lead the chain through and not take any more load itself.

Unlay one strand for rather greater length than will be used in the eye. Bend the remaining two strands into the loop through the chain, and tuck one end through the standing part (Fig. 20A). Take the other strand that has formed the loop

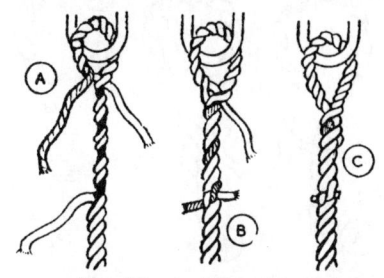

FIG. 20—A chain splice

and put it in the space left by the separate unlaid strand. Take it some way along the rope and join the two ends in one of the ways described for a long splice (Fig. 20B). This leaves the end that was passed between two strands. Taper it and tuck it against the lay two or three times (Fig. 20C). This splice is liable to get distorted, so it will benefit from rolling and stretching before the ends are cut off.

Grommet splice.

The grommet splice is also called an artificial eye. It makes an eye comparable to the basic eye splice or some of its variations, but it is without some of the bulk of the tucked part of the neck of those splices. When properly made its strength may not be much less than that of tucked splices. There is a family resemblance to the chain splice, but all three strands form the eye.

Unlay one strand for a distance rather more than is needed to form the eye and bend the remaining two strands into the loop (Fig. 21A).

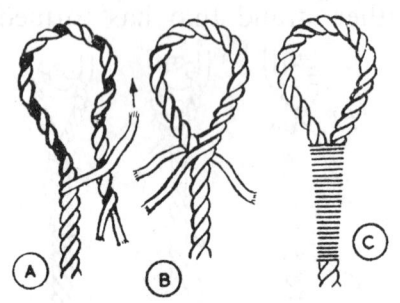

FIG. 21—A grommet splice

Try to avoid disturbing the lay and natural kinks of the loop or the separated strand. Lay up the single strand in its original place between the other strands, but going the opposite way round (Fig. 21B).

Separate the yarns or groups of filaments of each strand. Worm them around the spaces between the strands of the rope. Taper them in their length. Use some thin thread to half hitch along the rope to hold these worming ends in place, then serve over with stouter line from close to the eye (Fig. 21C).

That is the normal method of finishing a grommet splice or artificial eye. An alternative is to taper the ends and tuck them over and under against the lay or around main strands as in a sailmaker's splice. This is bulkier and then becomes alternative to the basic eye splice and not usually as acceptable.

If the splice is made in four-strand rope, start by unlaying one strand, as for three-strand. Make the loop and lay this in place the opposite way round. Having done that, find the strand opposite to it in the rope formation and unlay that from the loop and put it back in the opposite way. This leaves the splice with two strands going each way. Complete the splice in one of the ways suggested for three strands.

Flemish eye.

The Flemish eye is also called a spindle eye and it may share the name artificial eye with the previous splice. The particular use of the Flemish eye is in forming a very small eye, which need not be much further across than the diameter of the rope. It differs from other splices in this chapter in being equally suitable for stranded or braided rope. Construction is tedious and a test of a capacity for taking care.

To make the eye, at a suitable distance from the end whip the rope, then separate the strands between there and the end into individual yarns. Use a piece of round rod or tube to gauge the size of the eye. The inside measurement of the eye will be the same as the diameter of the rod. To assist in controlling the splice as it is formed, two humps may be built up each side of the centre, by binding around the rod with any available light line. While this arrangement is definitely a help, the humps can be dispensed with if care is taken.

Fasten short lengths of yarn along the rod at fairly close intervals (Fig. 22A). These will later be used as stops. Separate the yarns of the rope

FIG. 22—A Flemish eye

into two equal groups. Take a pair of yarns and join them over the rod with an overhand knot (Fig. 22B). Repeat with all the yarns, staggering the knots as much as possible, so as to avoid forming lumps.

Unfasten the yarns laid along the rod and use them to draw the knotted yarns into a neat ring, then withdraw the rod. The eye must now be tightly served over, starting at the top and working down each side to the neck. The ends of the yarns hanging down must be tapered and laid over the rope; straight if a braided rope, or wormed around if a laid rope. They are then served over to complete the splice (Fig. 22C). If it is expected that the eye will have to stand hard wear it maybe further protected by cockscombing (Fig. 67).

CHAPTER 5

END-TO-END SPLICES

THERE are a number of conditions in which ropes have to be joined more or less permanently end to end, where the common short and long splices may not be suitable, because of differences between the two ropes, shortness of ropes, etc.

Shroud knots.

There are several ways of making a shroud knot for joining ropes, which, although called a knot, is in effect more of a splice, and certainly deserves a place in a book of splices because of its one great advantage—its economy in material; a shroud knot

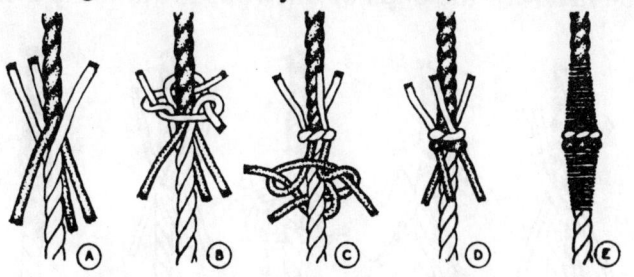

FIG. 23—A shroud knot (first variation)

using up less rope than any other joining knot or splice. It was originally used to repair a shroud (rope supporting the mast of a ship) after it had been shot away in action—a case where the

48   ROPE SPLICING

minimum shortening was desirable.

First variation—This is the commonest form of shroud knot. Unlay the ropes for a short distance and marry them (Fig. 23A). With the strands of one rope make a wall knot, in the opposite direction to the lay, around the other rope (Fig. 23B, also *see* Fig. 17). Pull the knot tight and do the same with the strands of the other rope (Fig. 23C). Pull both knots tight and close together (Fig. 23D). Taper the ends and worm them around the rope, then finish by serving over (Fig. 23E).

Second variation—This is made in a similar way to the first variation, but crown knots (Fig. 17) are used instead of wall knots. This variation is sometimes called French shroud knotting.

Unlay the ends and marry them (Fig. 24A). Form a crown knot in the opposite direction to the lay with the ends of one rope around the other

FIG. 24—A shroud knot (second variation)

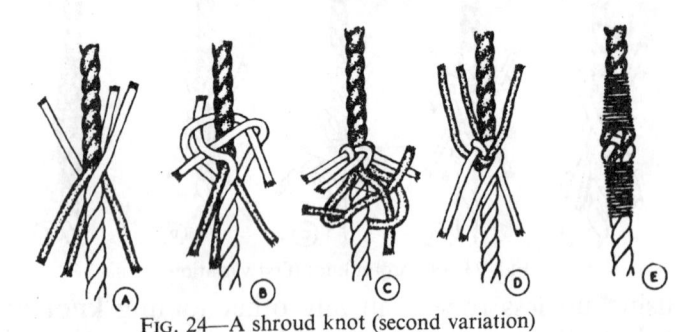

(Fig. 24B). Pull tight and repeat with the other ends (Fig. 24C). Pull the knots close together

ROPE SPLICING

(Fig. 24D) then taper the ends, worm them around the rope and serve over (Fig. 24E).

Third variation—This method uses a crown knot on one rope and a wall knot on the other, the strands at the finish all pointing one way. This is also sometimes called a French shroud knot.

Unlay the ends and marry them (Fig. 25A). With the strands of one rope make a crown in the direction against the lay around the other rope

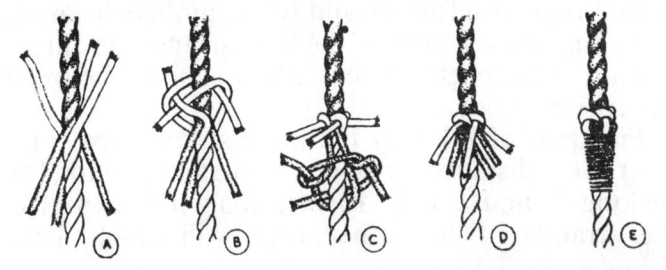

FIG. 25—A shroud knot (third variation)

(Fig. 25B). With the strands of the other rope make a wall knot against the lay (Fig. 25C). Pull the knots up close (Fig. 25D). then taper the ends well, lay them down over the rope and serve over (Fig. 25E).

Fourth variation—This is a development of the second variation. After forming the two crown

FIG. 26—A shroud knot (fourth variation)

knots and pulling them fairly closely together it will be seen that adjoining pairs of ends lie in opposite directions alongside each other. Take each end and follow it round alongside its partner until the whole knot is doubled, then cut the ends off (Fig. 26). The finished knot looks like a Turk's head or footrope knot.

Short splices (three-strand into four-strand).

Three-strand rope may be joined to four-strand in one of two ways; either by making the three-strand rope into four-strand for a suitable distance, or by making a length of the four-strand into three-strand. In strength there is little to choose between the two methods.

First method—Unlay the three-strand rope (Fig. 27A) for a distance sufficient to make the ends for tucking Y and a little further than it is expected the strands of the other rope will reach after

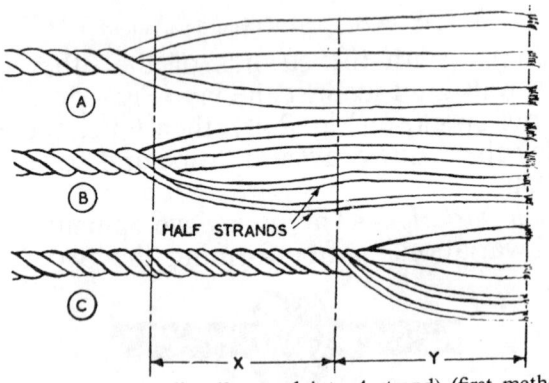

FIG. 27—A short splice (3 strand into 4 strand) (first method)

tucking X. Take one of the strands and divide its yarns into two equal parts, laying them up between the fingers to form two half-strands (Fig. 27B).

ROPE SPLICING

Take the two full thickness strands and the two half-thickness strands and lay them up together for the distance *X*, making a four-strand rope for that length (Fig. 27c).

Complete the splice in the same way as the common short splice, marrying the unlayed strands of the four-strand rope into the four ends of the three-strand rope, making two full and two taper tucks each way.

Second method—Unlay the end of the four-

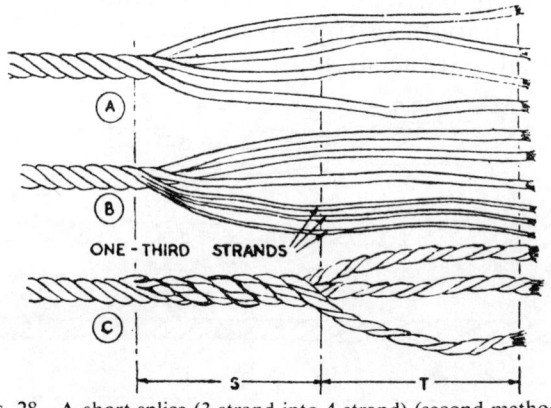

FIG. 28—A short splice (3 strand into 4 strand) (second method)

strand rope (Fig. 28A) for a distance sufficient for tucking *T* and a little further than it is anticipated the end strands of the other rope will reach when tucked *S*.

Divide the yarns of one strand into three equal parts (Fig. 28B) and twist them around the other three strands. Lay up the strands for the distance *S*, making a three-strand rope for that length (Fig. 28c).

Complete the splice in the same way as the common short splice, marrying the unlayed strands of the three-strand rope into the three ends of the four-strand rope and making two full and two taper tucks each way.

Long splice (three-strand into four-strand).

Most of the considerations affecting the common long splice in pairs of three- or four-strand rope apply to joining three-strand rope to four-strand rope. In particular the recommendations given for the amount of rope to be used according to its material and the strength required should be taken

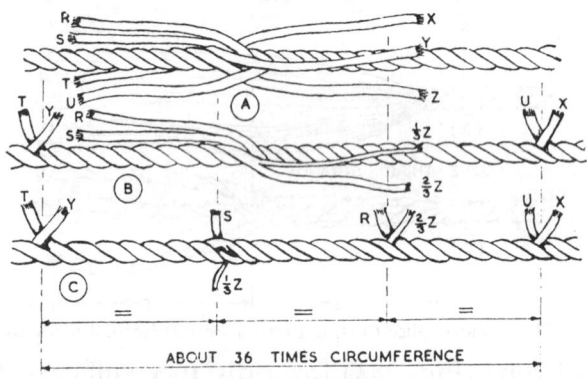

FIG. 29—A long splice (3 strand into 4 strand)

into account, using the figures as they apply to two four-strand ropes. Bring the two ends together, letting two ends of the four-strand rope come in one space of the three-strand rope (*R* and *S*, Fig. 29A).

Unlay one strand of the three-strand rope (Fig. 29B) and lay up one strand of the other rope in its place (*Y* and *T*). Do the same with another pair (*U* and *X*) in the opposite direction. By working each pair almost to the end of the laying up strand the junctions should finish about 36 circumferences apart (natural fibre rope).

Split the remaining strand of the three-strand rope *Z* into two unequal parts (roughly $\frac{2}{3}$ and $\frac{1}{3}$). Half-knot the thinner part to one of the remaining strands of the four-strand rope. Commence unlaying the last strand of the four-strand rope and lay up the thicker part of strand *Z* in its place. Arrange the junctions so that they are equally-spaced (Fig. 29c). Secure the ends by any of the methods described for the common long splice.

Alternatively, in place of the actions in the last paragraph, keep strand *Z* the full thickness and lay it up in place of *R* to the same point as shown for $\frac{2}{3}Z$. Finally tuck *S* where it is and secure all the ends by any of the methods described for the common long splice.

Short splice, knotted.

This differs from the common short splice only in the treatment of the strands when marrying the

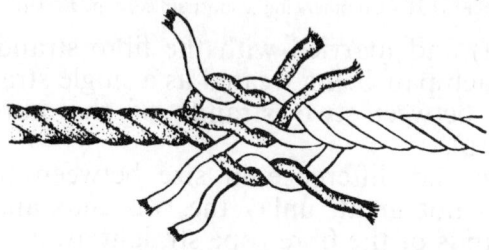

Fig. 30—Commencing a short splice, knotted

ends. Instead of the plain marrying, each end strand is twisted with its opposite number into a half knot. In the case of right-handed rope each strand is mated with the opposing one on its right (Fig. 30). When the knots are pulled tight the ends lie in position ready for tucking.

While this method of starting a short splice has the advantage of holding the parts together when starting it is generally considered to result in a weaker splice than the conventional method of starting.

Short splice (rope to wire).

There are a number of slight variations in making this splice, but in all methods the six strands of the wire rope are paired (the heart is

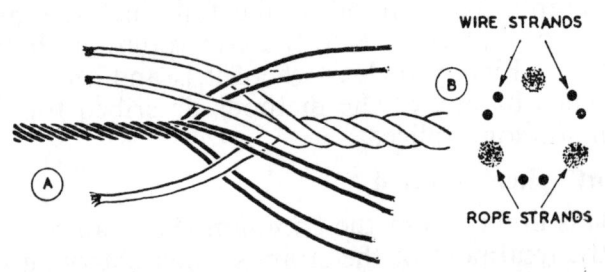

FIG. 31—Commencing a short splice, rope to wire

cut out) and married with the fibre strands (Fig. 31A), each pair being treated as a single strand and placed between two strands of the fibre rope (Fig. 31B).

When the difference in size between the two ropes is not great, unlay the two ends and taper the strands of the fibre rope straight away. Marry

ROPE SPLICING

the strands—two of wire to one of fibre—and tuck the wire 'over and under one' into the fibre rope. Tuck the fibre rope strands into the wire 'over and under two'. Three tucks each way are usually sufficient. Taper the last wire tuck by reducing to one strand.

If the wire is much thinner than the fibre rope difficulty will be found in tucking the fibre into the wire. Instead, unlay the fibre rope for some way (say 30cm or 12in), taper the strands fully for the whole unlaid length and lay up the strands to within a short distance of the end. Marry the wire and fibre, letting the unlaid wire strands be considerably longer than the unlaid fibre (60cm or 2ft is not too much, if the greatest strength is required).

Tuck the pairs of wire strands 'over and under one' into the fibre rope for as far as they will go, tapering by reducing to one strand towards the end. Cut off the ends of the fibre strands.

In both cases the best finish is obtained, if the wire is suitable, by separating the wires at the ends of each strand and sewing each wire, with a sailmaker's needle, into the fibre rope. For the best finish serve tightly over the whole splice.

Long splice (rope to wire).

This splice is more suitable than the short splices for cases where the fibre rope is considerably thicker than the wire, say upwards of four times greater.

Unlay the fibre rope for rather more than would be needed for a common short splice. Taper the strands. Unlay the wire rope for a considerably longer distance (100 to 150 times its own diameter). Cut out the heart. Take three alternate strands

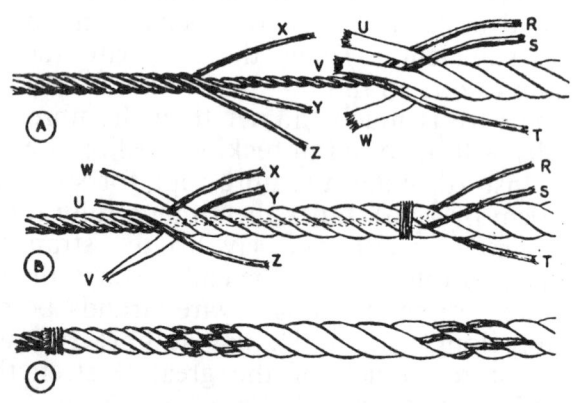

FIG. 32—A long splice (rope to wire)

and lay them up to form a three-strand wire rope. Marry the ends of this rope with the end of the fibre rope (Fig. 32A), letting strands *R*, *S* and *T* alternate with strands *U*, *V* and *W*. Put a temporary seizing over the junction.

Lay up the fibre strands around the thin wire rope as far as the other three wire strands, letting strands *U*, *V* and *W* finish up alternating with strands *X*, *Y* and *Z* (Fig. 32B).

NOTE—The two marryings should be much further apart than shown in the figure, where they are closed up for convenience in illustrating.

Remove the seizing. At both marryings tuck the wire around the fibre strands with the lay in the same manner as for a sailmaker's splice, i.e. work each end around the strand it is already under four or five times. Cut off the wire ends, or sew them into the fibre strands, as described for the short splice, rope to wire.

Lay the remaining ends of the fibre strands around the wire, with the lay, finishing with a tight seizing (Fig. 32c). Finally, for the neatest finish, serve over the entire splice.

Grecian splice.

This splice is of more ornamental than practical value, although it was used for repairs in the days of hemp standing rigging, and as a decorative feature when seamen had time to fill. It is neat, but is time-consuming and uses up more rope than a shroud knot or a common short splice. If made in synthetic rope wax should be used on all parts and rubbed into the nettles as they are formed, otherwise there will be difficulty in making the filaments stay twisted together.

Unlay the ends of the ropes to be joined for nearly twice as much as would be needed to make a common short splice and put on seizings. Note which are the outside yarns in each strand at the seizings and unlay them back from the ends, removing approximately the outside half of each strand: then lay these yarns up into stout nettles between the finger and thumb. Make the same number of nettles on each rope. Turn the nettles back over the seizings, out of the way, and lay up the thinned strands for just sufficient distance to take one tuck of short splicing.

58 ROPE SPLICING

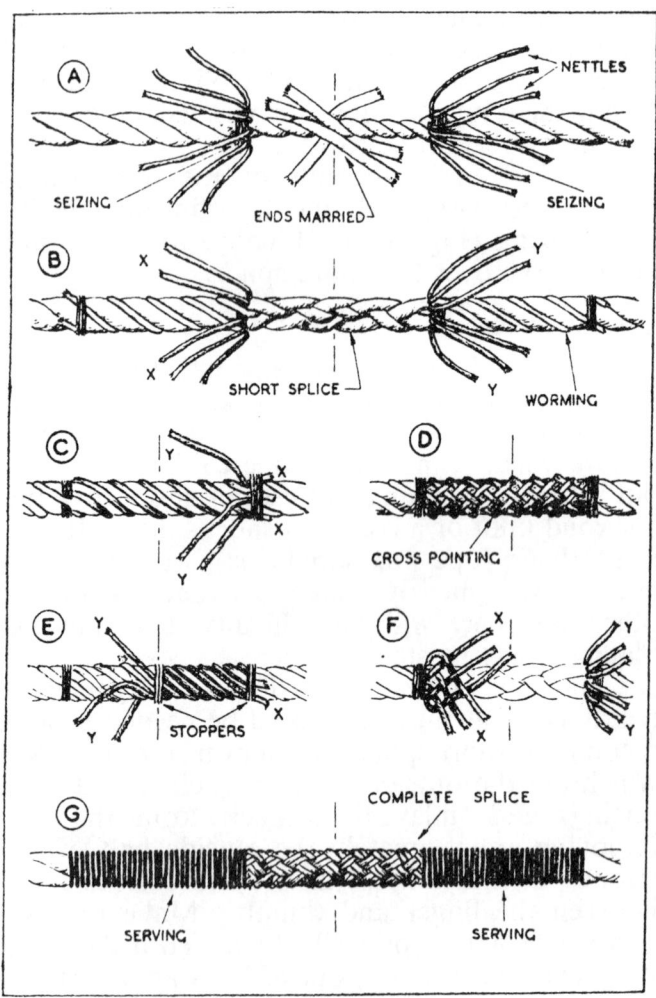

FIG. 33—A Grecian splice

ROPE SPLICING

Marry the ends (Fig. 33A) and make one tuck each way as in starting a common short splice. From each of the thinned strands take sufficient outside yarns to worm the rope and cut off the rest. Worm the rope each side for a distance about the same as the length of the short spliced centre (Fig. 33B).

The nettles, which have not so far been used, are next formed into cross-pointing over the short splice. There are several ways of doing this. The most straightforward method is: Take the nettles from one end X spirally around the rope, finishing up at the opposite end between the other nettles (Fig. 33C). The best angle of twist is about 45° to the centreline of the rope. Securely stopper their ends. Take the other nettles Y spirally, in the opposite direction back to the other end, tucking them 'over and under one' into the first nettles (Fig. 33D). The nettles may be worked singly, doubled or in threes, as preferred.

Alternatively, after laying nettles X spirally to the opposite end, lay nettles Y beside them. Put a temporary stopper at the middle of the splice (Fig. 33E). Now, treating the two sets of nettles in the same way as rope's strands when short splicing, tuck one lot 'over and under one' into the other lot from the stopper to the end of the splice. Get in as many tucks as possible. Remove the stopper and tuck the other nettles in the same way in the opposite direction.

The strongest, but most tedious method of cross-pointing the splice is as follows: With an even number of nettles at each end, start at one end X, using all the nettles, and take them alternately in opposite directions over and under each other (Fig. 33F). Continue this to the opposite end. At that end take nettles Y and tuck them back through the cross-pointing letting each nettle follow its opposite number back to the opposite end, so doubling the cross-pointing. With this method the correct degree of tightness for the first half of the cross-pointing can only be found by experience.

To complete the Grecian splice, after cross-pointing over the short splice, taper the ends of the nettles, lay them down over the worming, then marl and serve over as far as the cross-pointing (Fig. 33G).

CHAPTER 6

SPLICING CABLES

CABLE-LAID ropes may be spliced in the same way as hawsers, but because of their greater bulk it is necessary to introduce various means of tapering to prevent the splice being ugly and clumsy. Besides these splices, in which the three hawsers forming the cable are treated in the same way as the strands of a hawser-laid rope, there are a few splices particularly adaptable to the special construction of a cable.

Ropemaker's eye

This splice is used to form an eye for holding

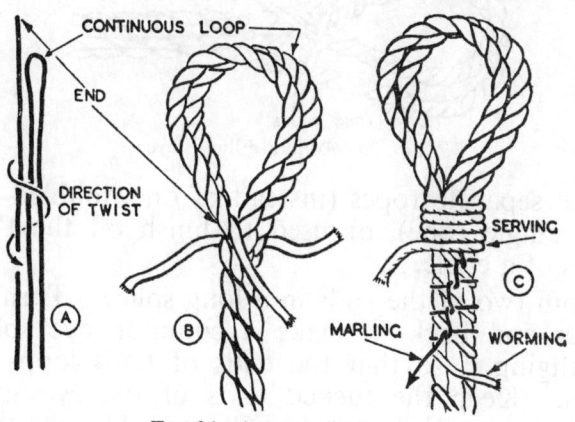

FIG. 34—A ropemaker's eye

the end when making a cable from three ropes. Two of the cable 'strands' are made from one doubled rope, while the other 'strand' is a single rope (Fig. 34A).

Form the end into an eye of the same size as the loop formed by the other two ropes (Fig. 34B). Separate the strands of the end and worm them round the cable. Secure this worming with marling (a series of half-hitches in light line, put on tightly, but not close together). Cover the worming and marling with a serving (Fig. 34C).

Admiral Elliott's eye.

This (Fig. 35), like the ropemaker's eye, is used for putting an eye in the end of a cable. It can be used when starting a cable to be made of

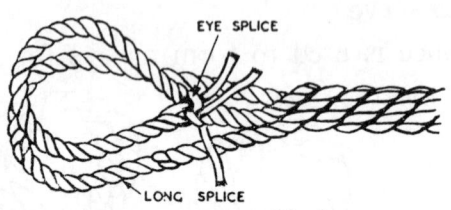

Fig. 35—Admiral Elliott's eye

three separate ropes (instead of one doubled and one single rope), or used to finish off the three ends of a cable.

Join two of the ends in a long splice. Turn the other end back to make a common eye splice, arranging it so that the ends of both loops are level. Keep the tucked ends of the eye splice fairly long and do not taper them. Use the three strands to worm around the cable, and complete the eye by marling and serving.

Common eye splice (cable).

To make a neat splice, unlay the three ropes for a sufficient distance from the ends to make the tucks and put on a stop. At about half of the distance from the end of each rope put on another

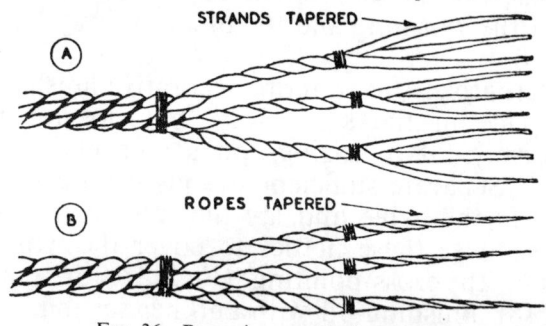

FIG. 36—Preparing a cable for splicing

stop. Separate the strands of each rope and taper them from the stop to the end (Fig. 36A). After tapering the strands lay them up again to form tapered ends to the ropes (Fig. 36B).

Take off the stops and use the tapered ends to tuck into the standing part of the cable to form an eye splice in the usual way (Fig. 3). The surplus ends may be cut off short, but a stronger and neater job will result if the ends are used to worm the cable and then marled and served over.

Short splice (cable).

Also called a drawing splice. Prepare the ends of both cables in the same way as described for the eye splice, by unlaying the ropes, separating the strands, tapering them and laying them up again. Marry the two cables and make a common short splice in the usual way (Fig. 8).

Use the tapered ends to worm around the cable, then tightly serve over the worming. Instead of serving, each worming may be held by three seizings, one at the end of the splice, one at the end of the worming and the other midway between the two.

Alternatively, for a more decorative finish, make the splice in the same way, but do not taper excessively. Put seizings on at the ends of the splice. Separate sufficient strands from each rope to worm the cable and lay the rest up into stout nettles. Use these nettles to cover the worming, either with cross-pointing (Fig. 33) or grafting (Fig. 18), finishing off with another seizing.

Long splice (cable). Sometimes called a mariner's splice.

This is in effect a two-fold-long splice—the rope forming the cable being laid up in each other's place as in a common long splice, then the strands

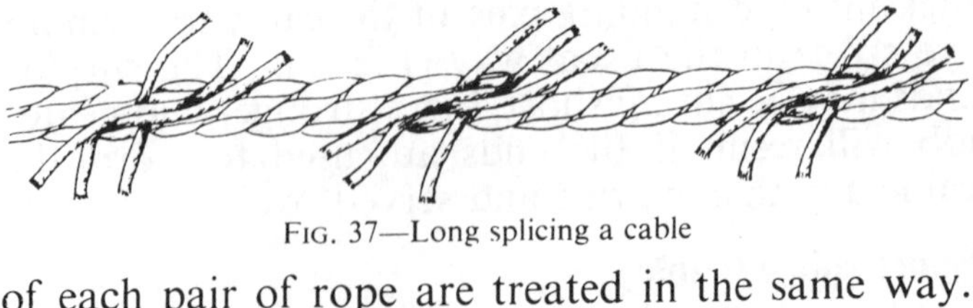

FIG. 37—Long splicing a cable

of each pair of rope are treated in the same way.

Commence by unlaying the rope and marrying them, treating them in the same way as the strands of a hawser in the common long splice (Fig. 9). Take a pair of opposing ropes, unlay one and lay the other up in its place. Do not take this as far as the common long splice about three times the circumference of the cable is sufficient. Do the same with another pair in the opposite direction. Marry the ends of the ropes (Fig. 37) and long splice them together. This is not easy and care must be taken to get each splice the correct tension so that when the cable is stretched all three ropes will be equally taut.

Staggered short splice (cable).

This is a compromise between the short and long splices, being less bulky than the short splice and less wasteful of material than the long splice.

Marry the ends of the cable, lay up and unlay pairs of ropes, then marry the ends of the ropes, as in the long splice (Fig. 37). Instead of long splicing the ropes together, short splice each pair, tapering the strands well after one tuck each way. It will be necessary to open the cable to expose the meetings of pairs of ropes to make their short splices. Great care is needed to see that each joint takes its fair share of the load when the cable comes under strain.

CHAPTER 7

BRAIDED ROPE EYE SPLICES

BRAIDED or plaited rope has many advantages in handling, easy running, keeping its shape and little tendency to kink. It presents special problems in splicing as different methods have to be used from the tucking of strands, which are simple and successful with rope made up of laid strands. Although most braided rope has an outside which is woven as a form of round plait, the inside construction varies. There may be a core of straight parallel yarns sufficient to fill the outer casing. There may be a second plaited cover inside the first and then the parallel yarn core. The inside may be a conventional three-strand rope with only the one plaited sheath.

The tightness and tension of braided rope, as well as the material of construction may affect the method of splicing chosen. Much traditional braided rope, made from natural fibres, had a very tight construction and methods of splicing that involved lifting yarns could not be used. Synthetic fibre-braided ropes are mostly polyester or nylon and of a much looser construction, so the casing and core can be manipulated to allow tucking and burying the working parts of a splice. A common form of exterior braiding is an eight-plait round

ROPE SPLICING

sennit, with each part formed by two, three or four yarns laid parallel to each other.

Tools (Fig. 38).

The knife and spike that are all the tools needed for many other splices should be supplemented with a few more tools for dealing with braided rope. It is necessary to take thread through the rope in some splices. This is best done with a sailmaker's needle, with a triangular point to force an opening to take the double thickness of stout thread. Needles are classed by gauge thickness. For the usual yacht ropes, up to about 20mm ($\frac{3}{4}$in), sizes between 12 and 17 gauge should be satisfactory. Although a few stitches might be made by forcing the needle with the side of a knife or against a bench, it is better to use a sailmaker's palm.

In several splices parts of the rope have to be pulled through. This might be done by attaching a thread and using a needle, possibly with the aid of a spike to make a sufficient opening and draw the part through. It may be possible to use a bodkin, as used for domestic needlework, for smaller ropes. Another tool that would do the job would be an awl with a needle eye or bodkin end. Some rug hooks or football lacing tools can be used. However, for the type of splice in which parts of the rope have to be buried for some

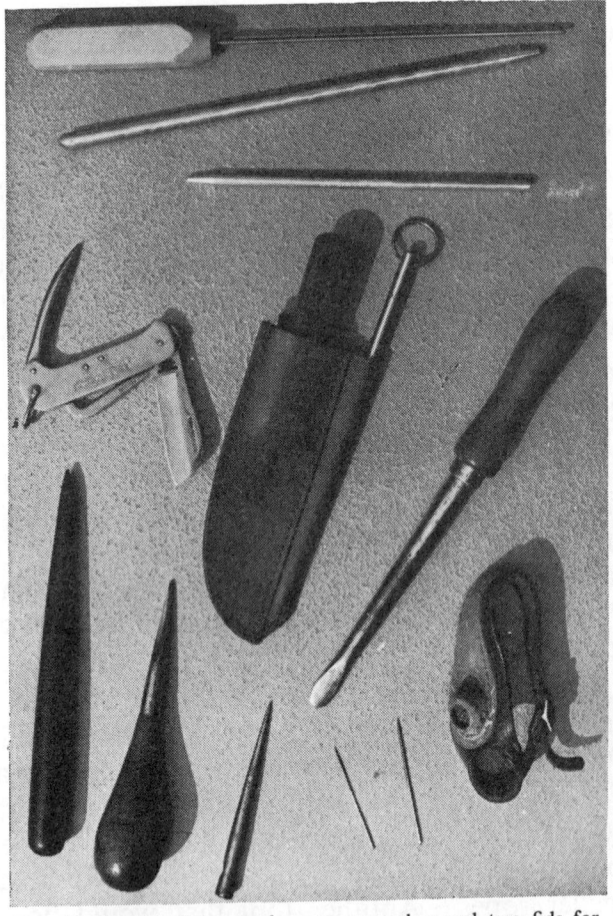

Fig. 38—Rope-splicing tools—top: a pusher and two fids for braided rope; centre: clasp and sheath knives with spikes, and a spike with flat end; bottom: a wooden fid and two marline spikes, two needles and a palm

distance, it is helpful to have special fids. These are solid rods with hollowed ends and recom mended sizes are shown (Fig. 39). With this goes a pusher, which is a handled rod about 3mm or

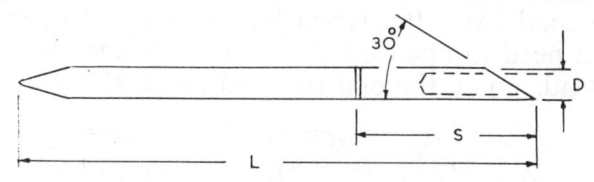

Rope diameter		Fid diameter (D)		Fid length (L)		Short section (S)	
mm	in	mm	in	mm	in	mm	in
6	$\frac{1}{4}$	5	$\frac{7}{32}$	140	$5\frac{1}{2}$	53	$2\frac{1}{16}$
8	$\frac{5}{16}$	6	$\frac{1}{4}$	172	$6\frac{3}{4}$	64	$2\frac{1}{2}$
9	$\frac{3}{8}$	8	$\frac{5}{16}$	198	$7\frac{3}{4}$	73	$2\frac{7}{8}$
10	$\frac{7}{16}$	9	$\frac{3}{8}$	242	$9\frac{1}{2}$	90	$3\frac{9}{16}$
12	$\frac{1}{2}$	10	$\frac{7}{16}$	280	11	103	$4\frac{1}{8}$
14	$\frac{9}{16}$	12	$\frac{1}{2}$	310	$12\frac{1}{4}$	108	$4\frac{1}{4}$
16	$\frac{5}{8}$	14	$\frac{9}{16}$	356	14	114	$4\frac{1}{2}$
18	$\frac{3}{4}$	17	$\frac{11}{16}$	406	16	121	$4\frac{3}{4}$
22	$\frac{7}{8}$	19	$\frac{13}{16}$	483	19	127	5
24	1	23	$\frac{15}{16}$	533	21	134	$5\frac{1}{4}$

FIG. 39—Fid sizes for splicing braided rope

4mm ($\frac{1}{8}$in or 3/16in) diameter. For thick braided ropes (over 25mm or 1in) a spring steel wire fid can be substituted for the solid one and pusher.

It is useful to have some adhesive tape (electrician's, masking tape or Sellotape) to use for temporary whippings. The rope has to be marked in some splicing. Wax crayon or felt-tipped pen are better than pencil or ball-point pen on synthetic ropes.

Flemish eyes.

The Flemish eye, described in Chapter 4, can

be used in braided rope, but the method of construction described there is best only when applied to a small eye. If a larger loop is to be used, there is no need to open up the strands around the eye. Instead, unlay the end for a distance about 15 to

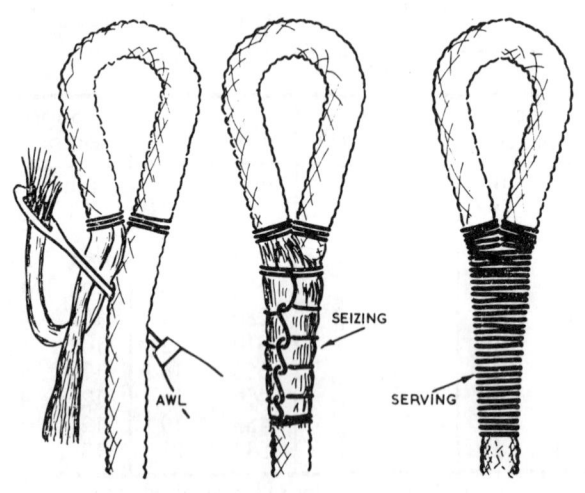

FIG. 40—Flemish Eye (braided rope)

18 times the diameter of the rope, so there are only a series of straight yarns. On some ropes temporary whippings may be advisable at the limits of the eye (Fig. 40).

Pass as many of these yarns, up to half the total, through the rope, using an awl or a special fid. How many can be tucked depends on the tightness of the rope construction. Do this reasonably close to the point of the eye, but not necessarily up to it. Lay all of the yarns down the rope. Taper them and cut them to various lengths to achieve a

ROPE SPLICING

tapered form generally. Seize them in place with a few half hitches of light line, then serve over. Serving is best done with the splice tensioned moderately.

A stronger variation of this splice uses stitches through the rope. Besides allowing enough length for the eye, include enough rope to lie alongside the main part for a reasonable distance (10 times

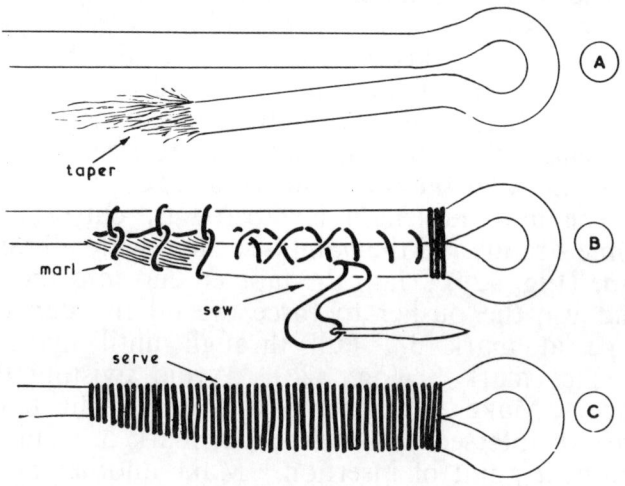

FIG. 41—Sewn and served eye in braided rope

the diameter would be about right). Beyond this allow about half as much again, which is unlaid and the yarns or filaments tapered (Fig. 41A). Draw the neck of the eye together with a seizing or West Country whipping (Fig. 41B).

Use doubled sail twine and a needle to sew the parts of the rope together, using large diagonal stitches through the centres of the parts of rope.

Spread the tapered yarns around the main part of the rope and seize them there. Put the splice under tension and serve over, from theeye to beyond the ends of the tapered yarns (Fig. 41c).

Lock tuck splice.

Where the braided rope has a single braided exterior around a core of straight yarns and the construction is sufficiently loose, it is possible to use a system of lock tucking followed by burying the end. A marked fid and pusher have to be used. Mark the end of the rope in two fid length increments. Bend the loop at the second mark from the end. With the loop adjusted to size, make a third mark at the other side (Fig. 42a).

The first fid length is tapered by cutting out some strands and the point is secured with adhesive tape (Fig. 42b). Jam the tapered end into the fid and use the pusher to force the fid through the rope at mark 3. Pull through until mark 2 reaches mark 3 (Fig. 42c). Avoid twisting the rope. Make another mark, four, of the braid pattern intersections away from mark 3. This is the next point of insertion. Make another mark $1\frac{1}{4}$ fid lengths further along from that.

Push the end of the fid through the braiding at the intersection mark and use the pusher to pass it along to mark 4 (Fig. 42d). Pull the end hard to tighten the locking tuck. Stroke the casing if it has become bunched, but do not stretch it yet. Cut off the end close to where it emerges. Stroke ('milk') the casing away from the eye until the cut end disappears and the braided pattern where

ROPE SPLICING

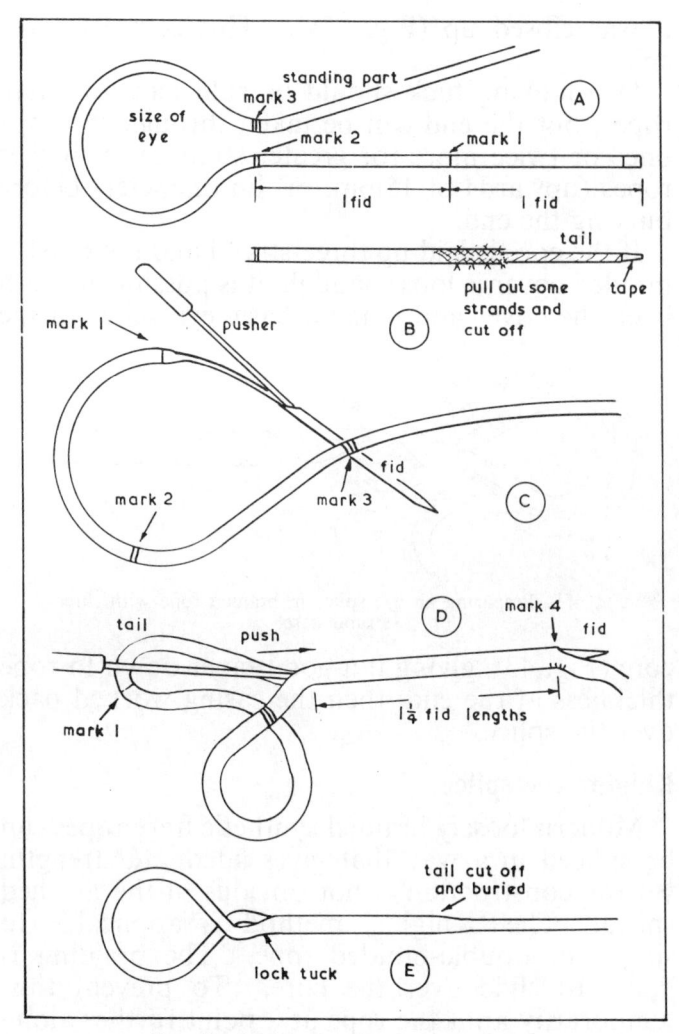

Fig. 42—Lock tuck splice in braided rope

it was closed up (Fig. 42E). This completes the splice.

One locking tuck should be sufficient for small ropes, but the end can be taken through the rope once or twice more for greater strength in thicker ropes (upwards of 15mm, or ⅝in diameter) before burying the end.

If the core is laid up three-strand and the outside braided cover is loose enough, it is possible to push back the cover and make a basic eye splice in the

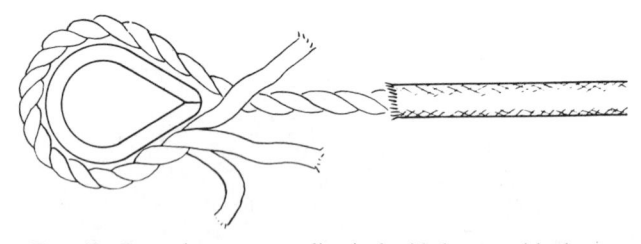

FIG. 43—Preparing an eye splice in braided rope with three-strand core

core (Fig. 43), giving it a good taper down to rope thickness at the end, then the casing worked back over the splice.

Knights eye splice

Modern loosely braided synthetic fibre ropes can be spliced in a way that gives adequate strength, yet the construction is not obvious in the finished splice. The 'Knights' method is applicable to single- or double-braided rope. The braiding is liable to slide over the core. To prevent this, temporarily knot the rope at a point further along the rope than the splice will reach (Fig. 44A).

ROPE SPLICING

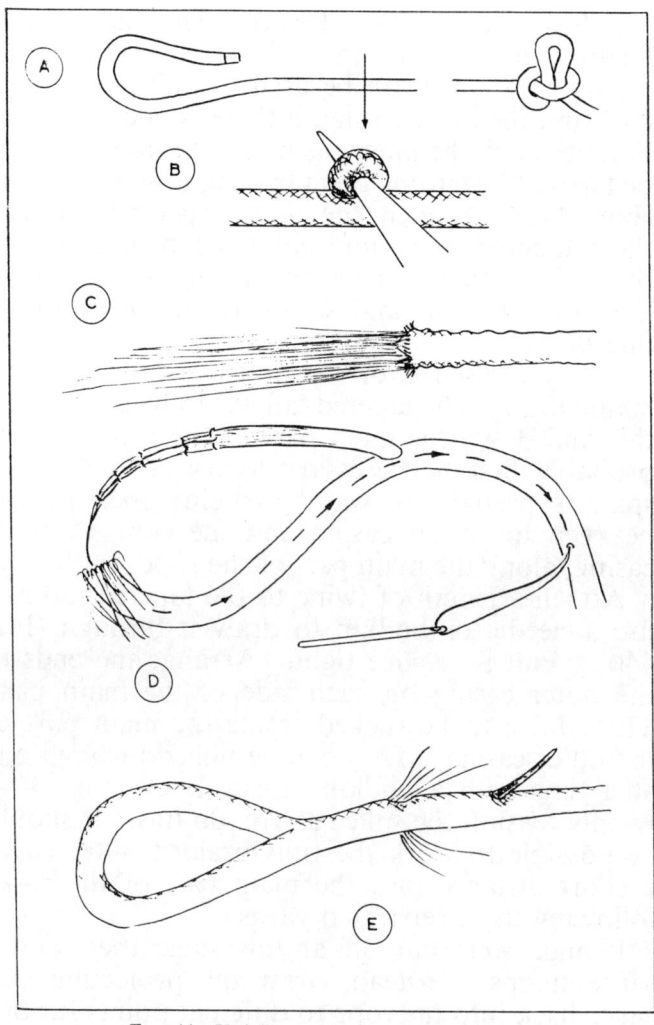

FIG. 44—Knights eye splice in braided rope

Allow enough rope for the eye and a short length for tucking, then force open the outer casing so a spike can be used to pull through the core and the inner casing, if there is one (Fig. 44B). Smooth back the inner casing so the core yarns can be tapered by cutting them back to different lengths (Fig. 44C). Stretch the inner casing back over these tapered yarns and either whip or tape it at the end. If there is no inner casing, wax and twist the tapered yarns and secure them with turns of fine thread.

Unlay the yarns of the outer casing for about 80mm (3in). The tapered tail has to be led through this and down the main part of the rope. It will probably first be necessary to use a long fid or spike to prepare the way by forcing open a space between the outer casing and the core, or inner casing, along the main part of the rope.

Attach a length of twine to the tapered tail and use a needle or bodkin to draw it through (Fig. 44D). Pull the splice tight. Arrange the ends of the outer casing on each side of the main part. These have to be tucked inside the main part of the outer casing. They can be pulled through one at a time with a bodkin or rug hook (Fig. 44E). Evenly tension the splice as you do this. It should be possible to work the fully-braided outer cover a short distance into the main part of the rope, following these separated yarns.

If ends were cut off at this stage they would cause lumps. Instead, draw the projecting end yarns back into the rope to different points further along the rope. Tension the splice and cut off all

ROPE SPLICING

projecting ends. Stroke the rope from the eye so as to spread the casing and lose the ends in the body of the rope, to finish with a neat appearance.

Samson eye splice.

Double-braided rope may be given an extremely neat and very strong splice. Allow a fid length from the end to a mark, then allow enough for the

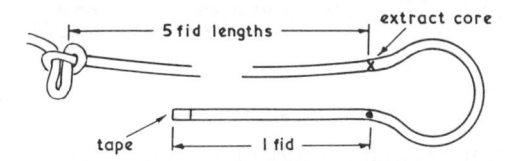

Fig. 45—Preparing an eye splice in double-braided rope

eye to another mark, and about five fid lengths further tie a knot to prevent the casings slipping (Fig. 45). A spike pushed through at that point will serve the same purpose.

At the second mark X open the cover and use a

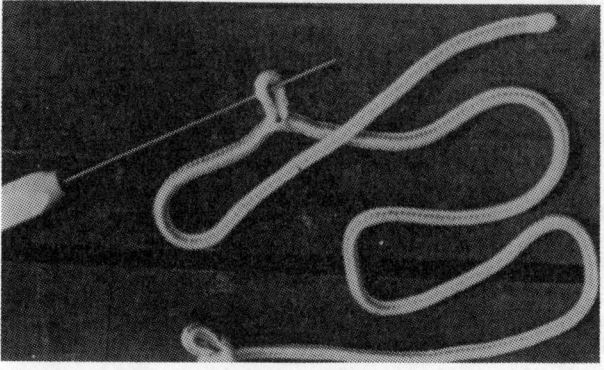

Fig. 46—Pulling out the core of a double-braided rope, to start an eye splice

spike to pull out the inner casing and core (Fig. 46). Smooth the cover from near the knot to check there has been no slipping. Slide the cover back to expose about ⅓ fid length and mark the core

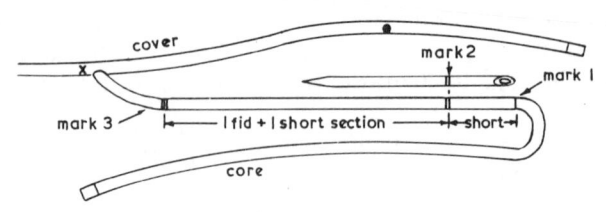

Fig. 47—Marking the core for an eye splice in double-braided rope

there, then expose a section 1⅓ fids long and mark it. Push the fid through the core at mark 2 (Fig. 47).

Pass the fid through the core so it just pokes out at mark 3, by bunching up or 'milking' the cover behind it. Put the end of the outer casing in the

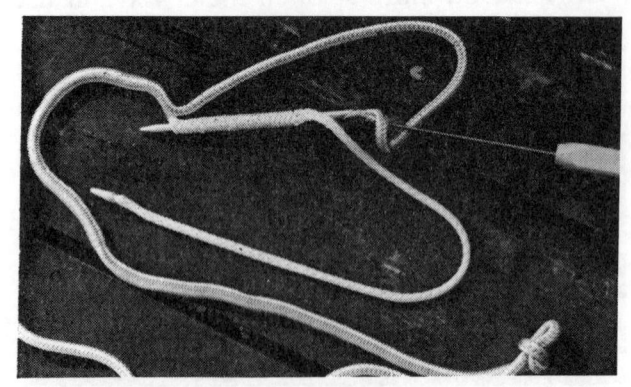

Fig. 48—Using the fid and pusher to pass the cover through the core in making an eye splice in braided rope

ROPE SPLICING

hole in the fid and jam the end of the pusher into it. (Fig. 48). Push the fid through until the first mark on the outer cover almost disappears. Remove the tools and leave the end projecting.

The next step is pushing the core end through the cover. Enter the fid at the first mark in the cover and pass it through to exit at the mark

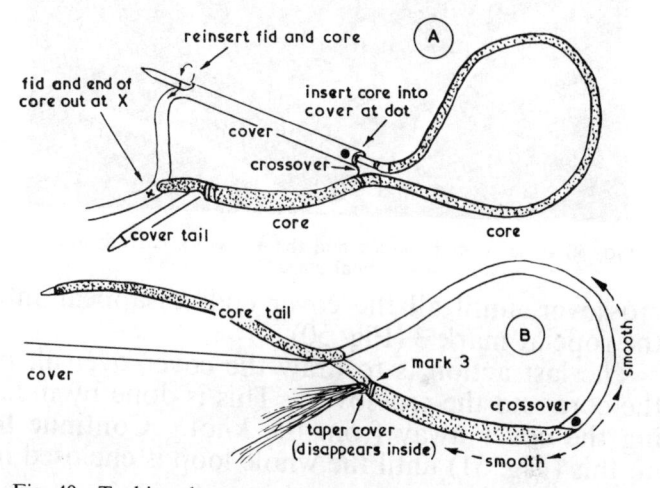

Fig. 49—Tucking the parts of an eye splice in double-braided rope

indicating the length of the loop. If the distance around the loop is more than the length of the fid, take the fid out at a convenient spot, pull through, then re-insert it (Fig. 49A). Pull the core end through until it is tight. Adjust the tension in the cover end so the crossover is tight in both directions.

The end of the cover has to be made to disappear inside the loop. Unlay a short length and cut to staggered lengths to achieve a taper (Fig. 49B). Smooth both sides of the loop away from the

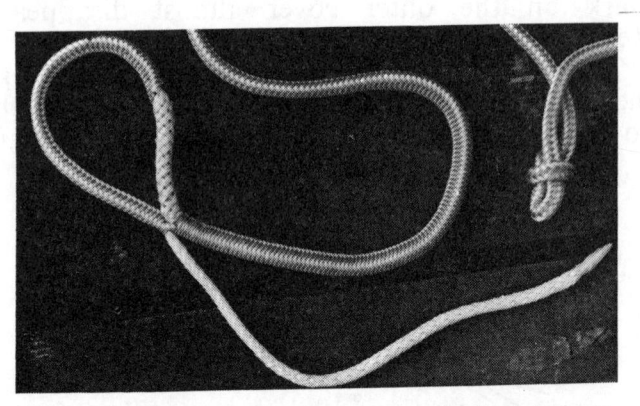

FIG. 50—The cover is buried and the eye splice is ready for the final stage

crossover, until all the cover ends disappear into the rope at mark 3 (Fig. 50).

The last action is to draw the cover over all of the core and the crossover. This is done by milking the cover away from the knot. Continue to do this (Fig. 51) until the whole loop is enclosed in

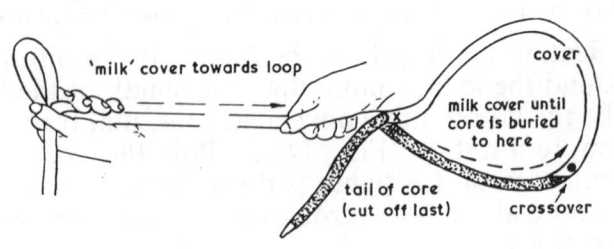

FIG. 51—Drawing the cover over an eye splice in double-braided rope

the outer cover. Smooth the eye towards the

ROPE SPLICING

core end. Cut off the tail reasonably close to the cover, but leaving a little exposed. Pull at the top of the eye and this end should disappear. Untie

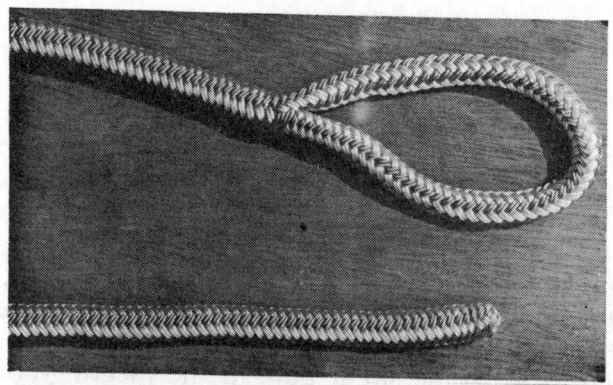

FIG. 52—Complete eye and back splices in double-braided rope

the knot and even up the braiding by smoothing and stretching the whole rope (Fig. 52).

If the splice is to be made around a thimble, this is inserted after the two parts are fitted into each other and before the eye is drawn to size. Splicing in new rope should present no trouble if the instructions are followed step-by-step, but if an eye has to be made in used rope, some shrinkage may have occurred. If the rope is soaked in water for a few minutes immediately before splicing, that will loosen and lubricate the fibres.

CHAPTER 8

OTHER BRAIDED ROPE SPLICES

As with twisted strand rope, eye splices are the main splicing need for plaited ropes. Ends are often better sealed and whipped than back spliced, but this is possible and less bulky than a back splice in twisted strand rope. There are no exact equivalents of short and long splices in braided rope, but satisfactory end to end splices can be made with some slight bulk at the meeting-point. When wire has to be joined to fibre-plaited rope, the splice may be neater than with twisted strand rope, while a joint between thick and thin ropes is also simpler and neater. It is advisable to master the making of a Samson eye splice first, as many of the other splices use very similar techniques.

Back splice.

This makes a neat end with little evidence of the splice having been made. Prevent the parts slipping on each other with a knot about five fid lengths from the end. Measure one fid length from the end and make a mark (Fig. 53A). Open the cover there and extract the core completely with a spike. Tape the ends of the core and cover. Mark the point on the core where it emerges, but before doing this stroke the cover from the knot

ROPE SPLICING

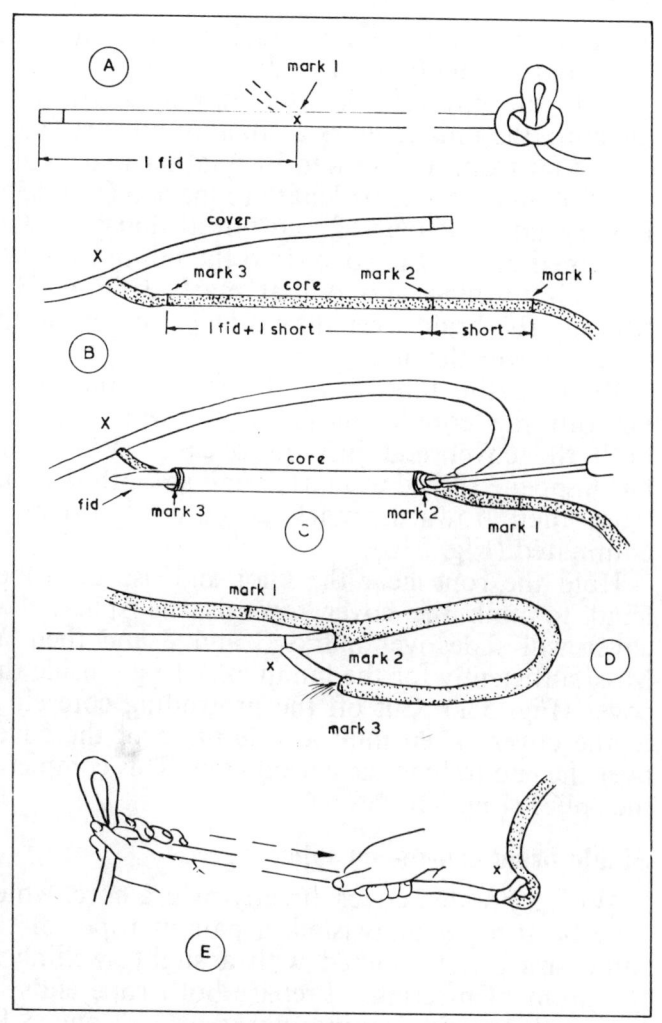

FIG. 53—Steps in making a back splice in double-braided rope

towards the end to remove slack and get inner and outer parts equally tensioned.

Slide the cover back after making mark 1 on the core, then make mark 2 equal to a short length on the fid from it, followed by another mark equal to a full fid plus a short length to mark 3 (Fig. 53B). Put the fid in at mark 2 and out at mark 3. Jam the taped end of the cover into the fid and use this to push through and out at mark 3 (Fig. 53C). Milking the braid over the fid helps in getting the fid and cover through.

Remove the tape from the end of the cover. Smooth the core from mark 2 towards mark 3 until the cover end just disappears. When this has happened, hold at mark 3 and smooth the core from there towards mark 2, until all excess is eliminated (Fig. 53D).

Hold the rope near the knot and use the other hand to milk the cover towards the splice. The cover will slide over marks 3 and 2 and then X. Milk sufficiently for the bump at X to go inside the cover (Fig. 53E). Cut off the protruding core close to the cover. Then milk a little more of the cover over the end to hide the cut-off core. This completes the splice (Fig. 52).

Single braid end-to-end splice.

With a braided cover directly over a core, which may be straight or twisted, a pair of ropes of the same size can be joined with a slight swelling at the point of meeting. Prepare both rope ends by taping them and making marks at 1, 2 and 3 fid lengths from the ends (Fig. 54A).

ROPE SPLICING

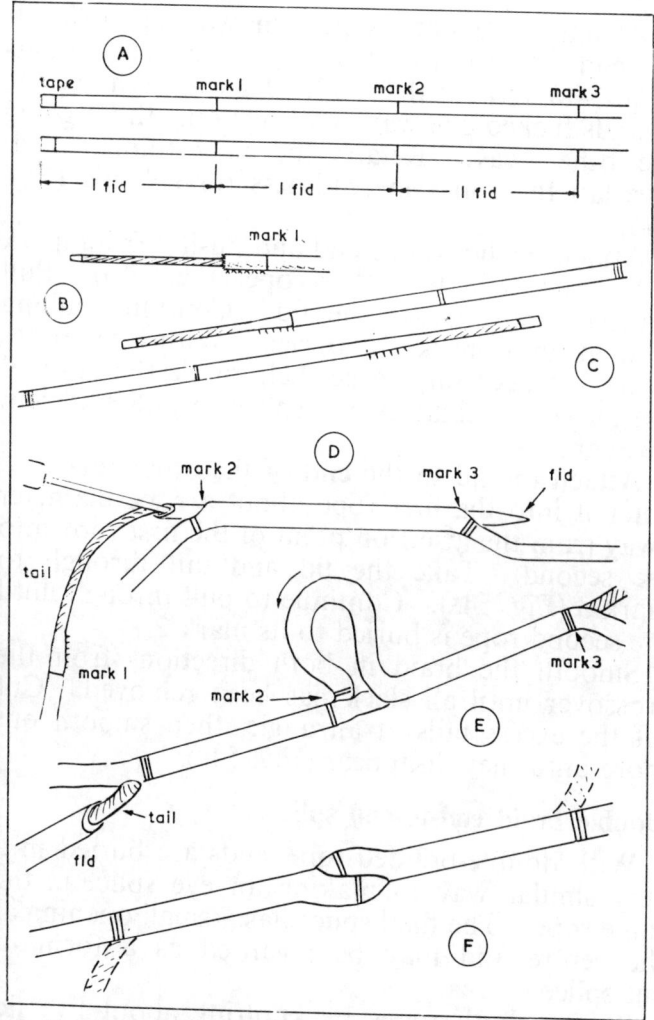

FIG. 54—Making an end-to-end splice in single-braided rope

At mark 1 count six pattern crossings towards the end. Pull out each group of marked strands (Fig. 54B) and cut them off. This will remove the strands tucked one way and leave only those going the other way. Retape the now thinner end. Overlap the ropes in readiness for splicing (Fig. 54C).

Attach the fid to one end and push it from mark 2 to mark 3 in the other rope (Fig. 54D). Pull through and remove the fid. Continue pulling through until mark 2 is buried. With large ropes it may be necessary to get help or tie the tail and use all your weight on the splice to pull sufficient through.

Attach the fid to the end of the other rope and enter it into the first rope about a rope diameter away from the insertion point of the first rope into the second. Take the fid and tail through to mark 3 (Fig. 54E). Continue to pull through until the second rope is buried to its mark 2.

Smooth the braid in both directions from the crossover until all slack has been removed. Cut off the excess tails at an angle, then smooth out more until they disappear (Fig. 54F).

Double braid end-to-end splice.

With double-braided rope, ends are buried in a very similar way to making an eye splice in the same rope. The final splice has a small opening at the centre and may be regarded as a form of cut splice.

Prepare both ropes by knotting about five fid lengths from their ends and taping the ends.

ROPE SPLICING

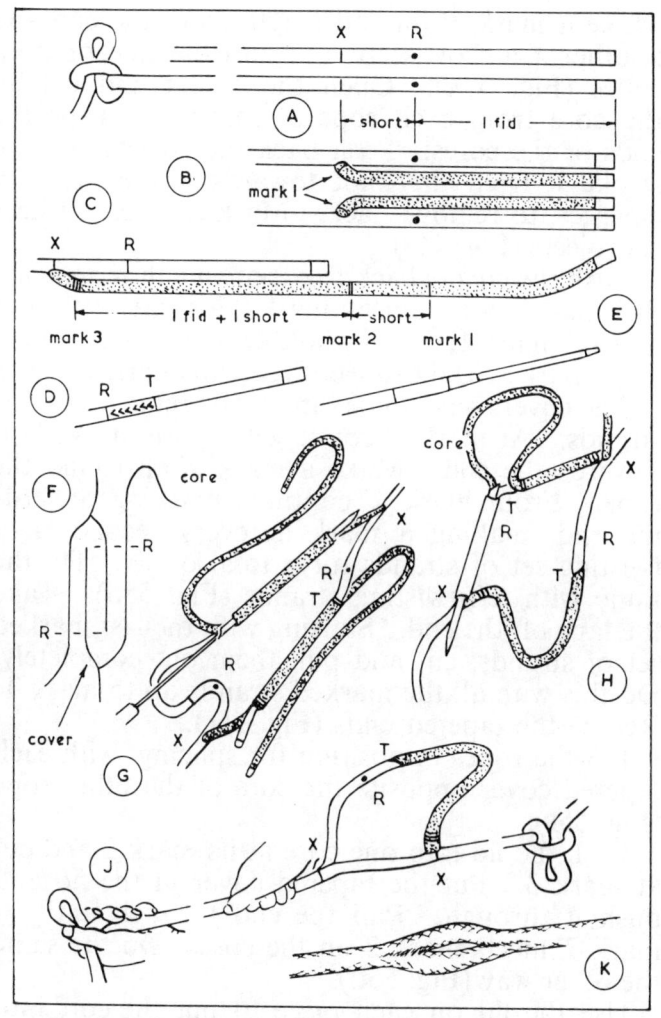

Fig. 55—Making an end-to-end splice in double-braided rope

Make a mark *R* one fid length from the end and another *X* a short section of fid from that on both ropes (Fig. 55A). Open the casing and extract the core from each rope at mark *X*. Tape the ends of the cores. Push back the cover and then stroke it from the knot towards where the core emerges, to remove slack. Mark each core where it projects (Fig. 44B).

Push the covers back to expose much more core, then make more marks: mark 2 a short section of fid from mark 1, then a whole and a short section of fid to mark 3. Make sure both cores match (Fig. 55C).

The covers have to be tapered by removing some strands. At mark *R* count seven pattern overlaps towards the end. Make a mark *T* all round the rope. From mark *T* continue marking towards the end, making a mark at every second right twisting set of strands for a total of six. Do the same with left-twisting strands (Fig. 55D). Take the tape off the end. Starting with the last marked set of strands, cut and pull them out completely. Do this with all the marked strands up to mark T. Retape the tapered ends (Fig. 55E).

Put the ropes in position for splicing, with each tapered cover opposite the core of the other rope (Fig. 55F).

Push the fid into one core at its mark 2 and cut at mark 3. Put the tapered cover in the fid and push it through. Pull the end through until its mark *T* meets mark 2 on the core. Do the same the other way (Fig. 55G).

Use the fid on each piece to put the core into the cover, entering at *T* and emerging at *X*. Pull through and get the crossovers tight by pulling opposing ends. Remove tape from the ends. Hold a crossover and stroke the braid smooth each side of it. The tapered cover should disappear Cut off each core close to mark X (Fig. 55H).

Hold each knot in turn and milk the cover towards the splice. The cover will slide over the surplus core. Continue until all slack cover has been taken up (Fig. 55J), leaving the small open centre as evidence of the splice (Fig. 55K). Untie the knots.

Joining thin and thick ropes.

In some rope applications it is an advantage to have a thin tail rope to pull through a thicker one. A neat, stront joint can be made, particularly if the smaller rope is about the same size as the core diameter of the larger rope.

Prepare the larger rope by knotting some way back from its end to prevent undue slipping of the

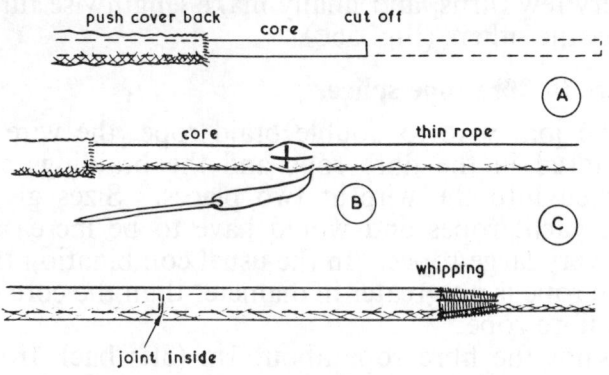

FIG. 56—Joining a smaller rope tail to a braided rope

cover over the core. Slide back the cover from the end and cut off the core at a distance equal to about 25 times the diameter of the rope (Fig. 56A). Bring the thin rope up to the core. The meeting ends may be fused together by using a flame while gently pushing the parts against each other, then with moistened thumb and finger they can be rolled round and parallel. This may be sufficient, but there can also be a few stitches over the joint, or the fusing can be omitted and reliance placed on stitches all round only (Fig. 56B). Smooth the cover over the joint and along the thinner rope. Put on a few turns of whipping line about three diameters from the end of the cover, then with this holding the strands, open the end of the cover and cut off some ends at irregular distances so as to taper the cover. Continue to whip over until past the cover strand ends.

It is advisable to put on the whipping with a needle. Take the thread through the rope after every few turns, and finally make lengthwise turns over the others (Fig. 56C).

Wire to fibre rope splice.

To join wire to double braid rope, the wire is enclosed in the fibre rope and the braidings are tucked into the wire at two places. Sizes given suit yacht ropes and would have to be increased for very large ropes. In the usual combination the wire rope is no greater in diameter than the core of the fibre rope.

Knot the fibre rope about 1m (3ft) back from the end. Slide back the cover from the end and

ROPE SPLICING

cut off 25cm (10in) of the core (Fig. 57A). Seal the end of the wire compactly, either by whipping or soldering. All wire strands can be cut the same

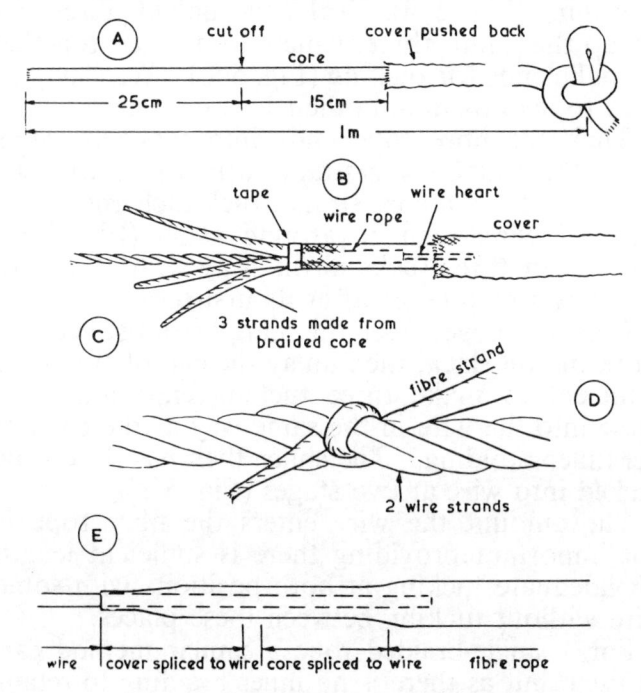

FIG. 57—Splicing wire and braided ropes

length, but if a tapered effect is wanted, some may be cut back a short distance or the core may be allowed to project. The wire will not be unlaid from the end in this splice, so sealing there is permanent.

Unlay some of the inner braiding ends sufficient for tucking into the wire (about 10cm, or 4in).

Push the wire into the core. If the fibre rope is too tightly laid, unlay more of the core to allow the wire to enter a reasonable distance. Tape or whip this joint (Fig. 57B). Take the unlaid fibres and group them into three, which are twisted together to make ends for tucking (Fig. 57C). Wax may be used to help them hold their twists.

Tuck the fibre rope ends into the wire rope. With the usual seven-strand wire rope, with six strands around a core strand, tuck each end under pairs of wire, then go around again (Fig. 57D). Do two or three tucks in this way with each end, which can be tapered after its first tuck.

Milk the cover over this part of the splice and work out all slack, then unlay the end of the cover sufficient to make three tucking ends and tuck these into the wire in the same way as the ends of the inner braiding. The splice then has fibre rope tucked into wire at two stages (Fig. 57E).

The amount the wire enters the fibre rope is not important providing there is sufficient length for adequate tucking at both positions with some wire without tucking between these places.

For a single-braided rope a similar method can be used, but as there is no inner braiding to retain the core in shape the first tucking has to be approached differently. If the core is laid as a three-strand rope, the joint may be made in the same way as described for a full three-strand fibre rope (Fig. 31). If the core has straight strands, they can be opened far enough and arranged around the wire rope, with tape or a whipping where the ends are to be divided into three groups for tucking.

CHAPTER 9

MISCELLANEOUS SPLICES

ROPE has been, and still is, constructed in ways other than laid in a twisted manner or braided. Some of these formations cannot be spliced satisfactorily, while others require special methods. There are other methods of ropework that use splicing principles although in themselves they may not be pure splices. There are other applications of splicing methods in special constructions, but once the standard splices are learned these adaptions are simple.

Eye splice (multi-plait rope).

This is a braided rope with eight strands, having the strands laid up from yarns, four right-handed and four left-handed. These are plaited in pairs (Fig. 58A). As the rope is made up loose and flexible, tucking is possible. Unlay a length about equal to five times the circumference. A temporary whipping at this point and where the strands are to be tucked, is advisable. Group the end strands into a pair right-handed and a pair left-handed for the front of the rope (Fig. 58B) and a similar pair for the back.

Tuck a right-handed pair against the lay under a right-handed pair of the standing part (Fig. 58C). Go in the same place and do the same with the left-handed ones (Fig. 58D). Turn the splice over and do the same with the remaining ends. Unlike

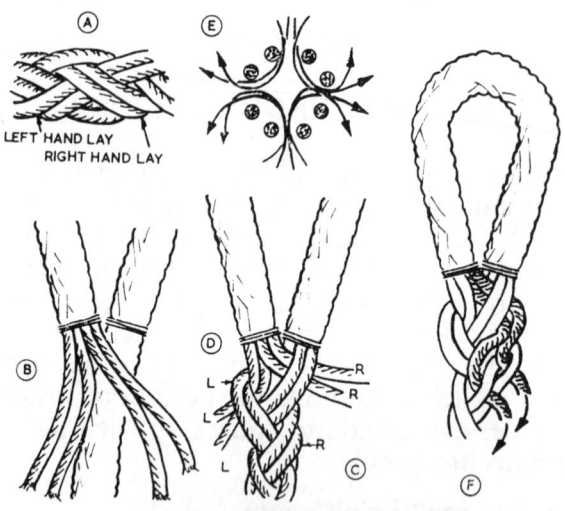

FIG. 58—Eye splice (multi-plait rope)

normal splicing, this will not result in an even spread of ends around the rope, but there will be two groups (Fig. 58E).

From this point the ends are tucked individually, but always right-handed under right-handed and the same the other way, working along the rope, each end going under one of a pair (Fig. 58F). Do this four times and dog-knot the ends together.

Two-ended eye splice.

There are occasions when two ends have to lead from an eye. Examples are the pair of jib sheets on a yacht and the corner guys on some tents. The splice used may be called a guy-line eye or a brummell splice. In three-strand rope, bend the rope into a bight and form the eye, then lift one strand near the eye and push the other half of

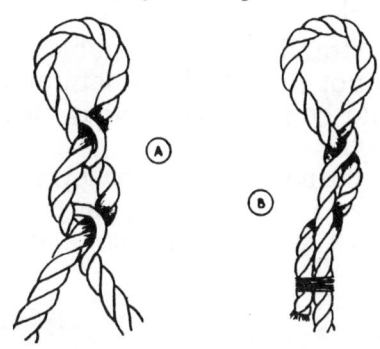

FIG. 59—A guy-line eye

the rope through. Lift a strand of the side of the rope that has been pushed through and bring the other side back through it (Fig. 59A). For most purposes there is an advantage, in use, in having the two tucks close together, but for strength let there be at least two strands between the tucks.

In braided rope use a fid and pusher for tucking, but the fid will have to be a size larger than used for other splices in that size rope, as here it has to take the whole rope and not just the cover or core. A shackle, snap hook or other fitting can be included in the splice, but this must be positioned before the first tuck.

The splice can also be used as a quick single-ended splice, possibly temporarily where a knot would be in appropriate. Use one short end and seize it to the main part after tucking (Fig. 59B). This version is sometimes made with the short end tucked twice, but it is stronger to have each side tucked once.

Cringle to sail.

Unlay a strand from a suitable size of rope, taking care not to disturb its natural shape. The length should be rather more than three times the finished length.

Pass the strand through one eyelet (Fig. 60A), keeping one end X twice as long as the other end Y. Twist the two strands together and pass end X

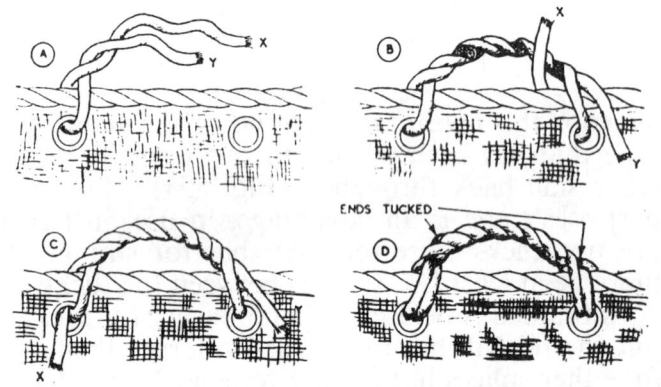

FIG. 60—Attaching a cringle to a sail

through the second eyelet (Fig. 60B). Lay it back around the cringle to the other end, forming a three-strand rope (Fig. 60C). Adjust the three parts to an even tension. Double the ends back on the cringle and tuck them into its strands, tapering for a neat finish (Fig. 60D).

For a stronger cringle it is possible to work around a further number of times. Any odd number may be finished off in the same way as for three strands. However, the larger number of strands do not lie together smoothly (due to the absence of a heart strand) and it is preferable to use a thicker strand worked three times only.

Another variation is to make the cringle as in Fig. 60, but instead of tucking the ends, lay them up around the cringle until they meet on its crown, making a four-stranded cringle. Finish the meeting ends off in the same way as a long splice.

Cringle to rope.

A cringle may be attached to the bolt rope of a sail or net in the same way as to the eyelets in a sail, taking the ends through the rope instead of

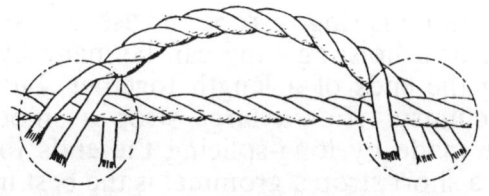

FIG. 61—One method of attaching a cringle to a rope

through the eyelets. Alternatively, a short length of rope may be cut to length and its ends spliced to the bolt rope (Fig. 61), in the same way as for a common eye or branch splice.

A better method is to use a single strand, tucking it under a single strand of the bolt rope (Fig. 62A), one end Y being long enough for tucking and the other end X being long enough for working the cringle. Double end X back around itself and under a strand at the same end as Y (Fig. 62B).

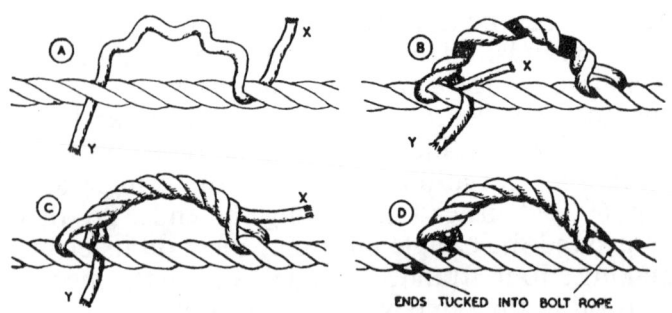

FIG. 62—A second method of attaching a cringle to a rop

Take it back to form the third strand (Fig. 62c). Even up the tension of the cringle and tuck both ends against the lay into the bolt rope (Fig. 62d).

Grommet.

A continuous ring of rope for use as a strop for a block or a hoisting sling can be made by short-splicing the ends of a length together, but this is rather clumsy. For a large strop a smooth job may be made by long-splicing the ends together, but for a short strop a grommet is the best method.

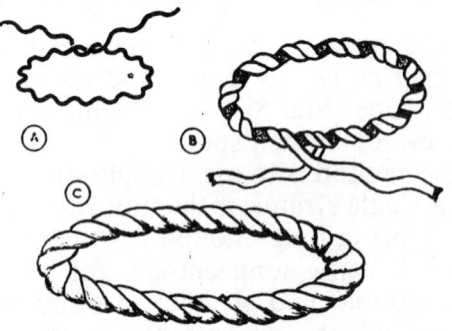

FIG. 63—Steps in making a grommet

ROPE SPLICING

Use a single strand with undisturbed lay, about $3\frac{1}{2}$ times the length of the final circumference of the strop. Bend it into the required size ring and commence twisting the ends together (Fig. 63A). Continue until they meet (Fig. 63B) then follow round once more with one end to make a three-strand rope and finish off the ends as in a long splice (Fig. 63C).

Selvagee strop.

A selvagee strop is more supple than either a grommet or a ring made by splicing a rope's ends together. For a given circumference it is also stronger. Although for convenience in illustrating

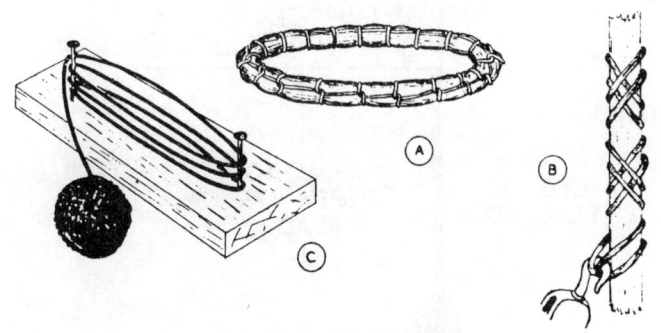

FIG. 64—A selvagee strop

the strop is shown comparatively thick for its size (Fig. 64A), it shows to best advantage as a long strop for such purposes as attaching a block to a rope or pole (Fig. 64B).

The strop is made of any suitable light line, such as spunyarn or marline. Drive two nails into a plank at a suitable distance apart. Attach the end

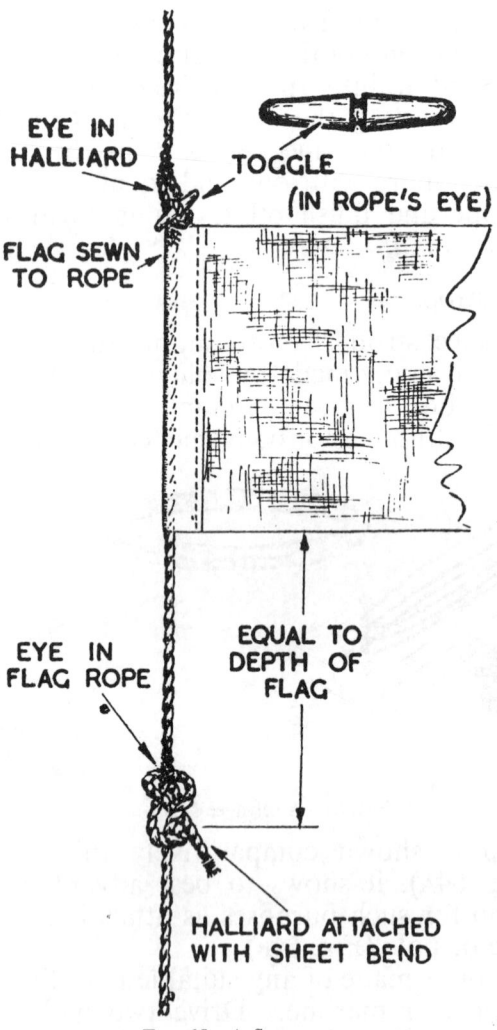

FIG. 65—A flag rope

of a length of line to one nail and wind on sufficient turns to make the strop (Fig. 64c). Secure the yarns together by marling all round (Fig. 64A).

Flag rope.

The rope attached to a flag should be made in the proportions shown in the sketch (Fig. 65). The toggle, or sny, at the top should be held tight in an eye splice and the flag sewn to the rope as close up to the toggle as possible, so that the flag may be hauled fully up the truck of a mast. The eye in the end of the halliard should be just large enough to push the toggle through end-on. Making the rope tail below the flag the same as the depth of the flag ensures the correct spacing when two flags are hoisted together.

Pudding splice.

When one strand of a rope is damaged by wear or accident it can be cut out and replaced by a new strand (Fig. 66). Cut out the damaged strand for a length of twelve times the circum-

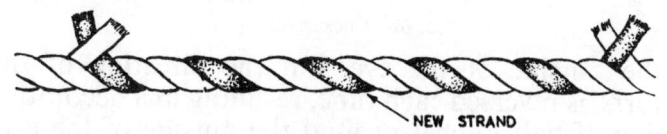

FIG. 66—A pudding splice

ference or more and carefully lay in the new strand in its place, preferably letting the new strand follow the other in as it is being removed so as not to disturb the other two strands. Join the ends of the old and new strands by any of the ways shown for finishing a long splice.

Cockscombing.

Where a spliced eye is not protected by a thimble, and is subject to friction, it has to be protected in some other way. Serving over tightly, as shown for the Flemish eye (Fig. 22) is satisfactory, except that if any part of the serving is cut, the whole serving will unwind. Half hitching (marling) the turns of serving will hold on to any broken ends, but as the half hitches have to be staggered some will come on the inside of the eye and because of their greater thickness will tend to wear first.

This can be avoided if cockscombing is used (Fig. 67). Each turn of serving is half hitched on

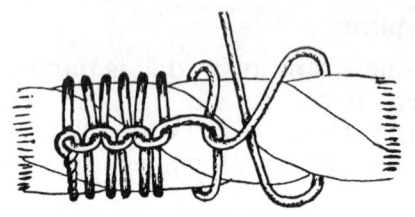

FIG. 67— Cockscombing

the outside of the eye, but the direction of the turns is reversed each time, resulting in a decorative line of half hitches around the outside of the eye.

GLOSSARY–INDEX

GLOSSARY—INDEX

	PAGE
Admiral Elliott's eye	62
Against the lay. In the opposite direction to that in which the strands of a rope lie	13
Artificial eye	45
Back splice, laid rope	24
Back splice, braided rope	83
Becket. A small loop in a rope	96
Bight. A loop	13
Block. A wood or metal device containing one or more pulley-wheels or sheaves	38
Bolt rope. The rope attached to the edge of a sail or net	34
Braided rope. Formed of untwisted yarns laid up around a heart in the form of cross-pointing	66
Branch splice	32
Brummel splice. A double-ended eye splice	95
Chain splice	42
Clove hitch. A jamming form of two half-hitches	39
Cockscombing. A method of serving a ring	102
Coir. Coconut fibre used for making rope which will float	10
Cotton. Used for making a flexible and soft rope	10
Cringle. A loop of rope attached to a loop of rope or a sail, usually around a thimble	96
Cross pointing. A form of round plait	60
Crown knot. Made by interweaving rope strands	25
Cut splice	32
Dacron. Trade name for polyester filaments	11
Dog knots	21
Draw. Tend to pull apart	24
End-to-end splices in braided rope	84
Eye in middle of rope	33
Eye splice. Any way of making a permanent eye in a rope	16
Eye splice, braided rope	72
Eye splice grafted	39
Eye splice, multi-plait rope	93
Eye splice with collar	38
Eye splice. wormed and collared	36
Fairleads. Metal channels to guide a rope and prevent chafe	42
Fibre. The fine strip of material used as the basis of the yarn of a rope	13
Fid. A wood or metal spike used to open a rope for tucking	14
Filament. Continuous synthetic fibre	11
Flag rope	101
Flemish eye	45
Flemish eye, braided rope	69

GLOSSARY-INDEX

	PAGE
Footrope knot	37
French shroud knot	49
German eye splice	41
Grafting. A decorative method of covering a rope with inter woven yarns	39
Grecian splice	57
Grommet. An endless rope ring	98
Grommet splice	43
Guy-line eye	94
Halliard, or halyard. A rope used for hoisting anything	101
Hawser-lay. The strands of a rope laid up so that when the rope is held vertical, the front strands slope upwards to the right	13
Heart strand. A straight hand laid along the centre of a rope when there are more than three outside strands	13
Hemp. Natural fibre, once commonly used for rope	10
Horse-shoe splice	33
Knights splice. Method of splicing braided rope	74
Lay. The direction in which the strands are twisted up to make a rope	13
Laid rope. The common type of rope, made by twisting strands in one direction only	13
Left-handed lay. The strands of a rope laid up so that when the rope is held vertical the front strand slopes upwards to the left	13
Locking tuck. Two end strands tucking in opposite directions around a main strand	41
Log line splice	32
Long splice	27
Long splice (rope to wire)	55
Long splice (three-strand into four-strand)	52
Manila. Fibre made from plantain leaves, once used for best ropes	10
Manrope knot	39
Marl. To bind with line in a series of half-hitches	102
Marline	99
Marline spike. A steel spike	14
Marry. To place the ends of rope together so that their strands interweave	26
Mariner's splice	64
Milking. Hand-smoothing a braided cover into place	80
Multi-plait rope. Rope construction with strands laid in both directions	93
Nettle. A thin line made by twisting up yarns between the fingers	39
Nylon. Synthetic material used for ropemaking filaments	11
Parcelling. Covering a rope with strips of canvas prior to serving over	32
Plaited rope. Alternative name for braided rope	66
Pointing. Working the end of a rope into neat ornamental taper	39
Polyester. Synthetic ropemaking material (Dacron, Terylene)	11
Polypropylene. Synthetic ropemaking material	11
Pudding splice	101
Right-handed lay. Same as hawser lay	13
Ropemaker's eye	61
Sailmaker's splice	34
Samson splice. Method of splicing double-braided rope	77

GLOSSARY-INDEX

	PAGE
Seize. Secure by binding with line	22
Selvagee strop	99
Serve. To cover a rope with a close binding of light line	33
Short splice	26
Short splice in braided rope	84
Short splice, knotted	53
Short splice (rope to wire)	54
Short splice (three-strand into four-strand)	50
Shroud. A side stay of a mast	47
Shroud knot	47
Shroud-laid rope. Four-strand rope, usually right-handed	13
Sisal. Natural material used for cheap ropes	10
Size of rope. Diameter is used in metric size, but circumference may be used for inches	12
Sny. A small toggle	101
Spike. A tapered wood or metal tool used for forcing rope strands apart	14
Spindle eye	45
Splice. Joining ropes by interweaving their strands	9
Spunyarn. A light line made from a number of yarns twisted up right-handed	99
Stop or stopper. A short seizing	26
Strand. The main member of a rope, made up of yarns twisted together	12
Strop. An endless rope	99
Tail rope. A thinner rope spliced to a thicker one	89
Tail wire. A wire rope spliced to a fibre rope	90
Terylene. Trade name for polyester filament used in ropes	11
Thimble. Metal or plastic eye to fit in eye splice to prevent wear	22
Wall knot	48
Warp. The nettles laid around the rope in grafting	39
Whipping. A short seizing, usually formed at the end of a rope to prevent it becoming unlaid	17
Wire rope. Rope made with metal strands, usually galvanised or stainless steel and usually six strands around a centre strand	54
With the lay. In the same direction as that in which the strands of a rope lie	13
Wormed, parcelled and served. The complete protection of a rope by filling the spaces between the strands, covering with canvas and binding over	33
Worming. Filling the spaces between the strands of a rope with yarns to make the surface of the rope circular, ready for serving over	36
Yarn. A number of fibres twisted together	13